Guide for the Total Solar Eclipse of 2045

USA Edition

Fred Espenak

Edition 1.0
September 2024

Guide for the Total Solar Eclipse of 2045 – USA Edition

Astropixels Publishing
P.O. Box 16197
Portal, AZ 85632

Astropixels Publishing Website: *www.astropixels.com/pubs*

This book may be ordered at: *www.astropixels.com/pubs/Guide2045.html*

Printed in the United States of America

ISBN 978-1-941983-53-9

Astropixels Publication: AP048 (Version 1.0a)

First Edition

Front Cover: Global map of the Aug 12, 2045 total solar eclipse through the USA, West Indies and South America.

Back Cover: Fred Espenak in Mazatlán, Mexico: Copyright ©2024 by Pat Espenak.

More about the eclipse can be found at:

www.eclipsewise.com/solar/SEnews/TSE2045.html

Table of Contents

Photo 1 – The Sun's corona is only visible to the naked eye for a few minutes during a total eclipse of the Sun. Total Solar Eclipse of March 29, 2006 from Jalu, Libya. © 2006 F. Espenak, www.MrEclipse.com

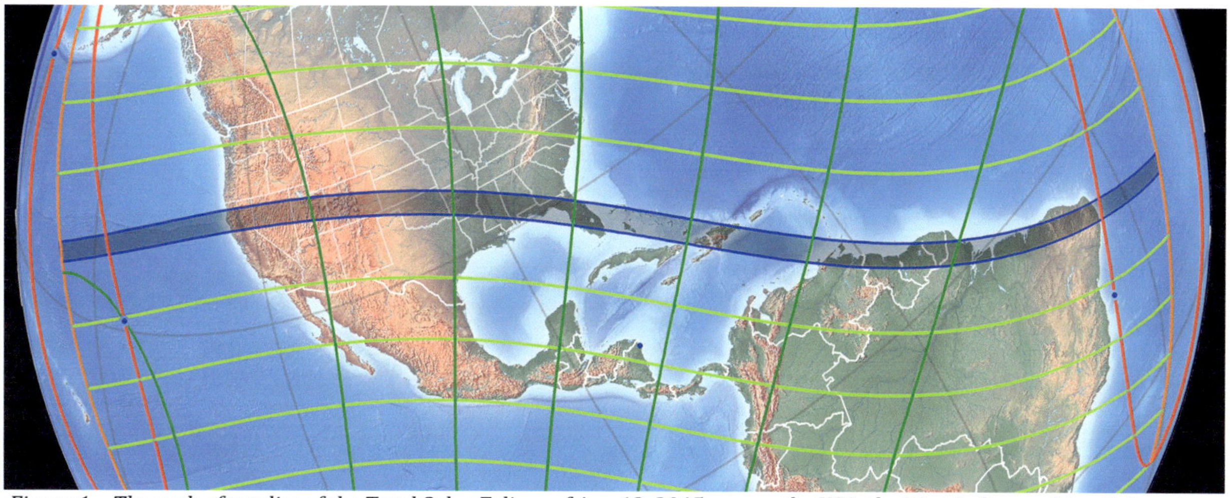

Figure 1 – The path of totality of the Total Solar Eclipse of Aug 12, 2045 crosses the USA, the West Indies, and South America.

1 – Introduction

On Saturday, August 12, 2045, a total eclipse of the Sun will be visible from the Western Hemisphere. Although a partial eclipse will be seen from most of North and South America, the total phase in which the Moon completely covers the Sun (known as totality) will only be seen from within the narrow path of the Moon's umbral shadow as it sweeps across eleven countries and territories: the United States[1], the Bahamas, the Turks and Caicos Islands, Haiti, Dominican Republic, Trinidad and Tobago, Venezuela, Guyana, Suriname, French Guiana, and Brazil.

During totality, the Sun's faint and exquisitely beautiful outer atmosphere – the solar corona – can be seen with the naked eye, the landscape is plunged into an eerie twilight, and brighter stars and planets appear. A total solar eclipse is one of the most awesome natural spectacles, but to witness it you must be in the *path of totality*.

The path of the Moon's umbral shadow begins in the North Pacific and travels east to reach the coast of California. The central duration of totality already exceeds four minutes as the path continues across Nevada and Utah. In Colorado, the duration is over five minutes. Onwards through Kansas, Oklahoma, Arkansas, and Mississippi, the duration surpasses the six-minute mark in Alabama and Florida.

Leaving the USA, the eclipse track crosses the West Indies including the Bahamas, Haiti, and the Dominican Republic. It reaches the Caribbean coast of Venezuela and hugs the northern coasts of Guyana and Suriname. After crossing French Guiana, the umbral shadow path sweeps through northern Brazil. The last three minutes of the total eclipse track brings the Moon's umbra across the Atlantic Ocean where it ends at sunset.

A partial eclipse will be visible from a much larger region including most of North and South America. Eclipse glasses or some type of eye protection must be used for viewing the partial eclipse. Of course, those in the path of the total eclipse must remove protective glasses to witness totality itself.

This guide is designed to assist observers in locating and reaching the central eclipse path to enjoy the awesome experience of totality.

---

[1] The path of the total solar eclipse of Aug 12, 2045 crosses parts of fifteen states: California, Nevada, Utah, Colorado, Kansas, Texas, New Mexico, Oklahoma, Missouri, Arkansas, Mississippi, Louisiana, Alabama, Georgia, and Florida. This is the first total eclipse to cross the USA from coast-to-coast (Pacific to Atlantic) since Aug 21, 2017 (14 states). The total eclipse of Apr 8, 2024 crossed the USA from Texas to Maine (15 states) plus Mexico and Canada. The upcoming total eclipse of Mar 30, 2033 crosses northern Alaska and Russia, while the eclipse of Aug 23, 2044 crosses North Dakota and Montana (plus Canada and Greenland).

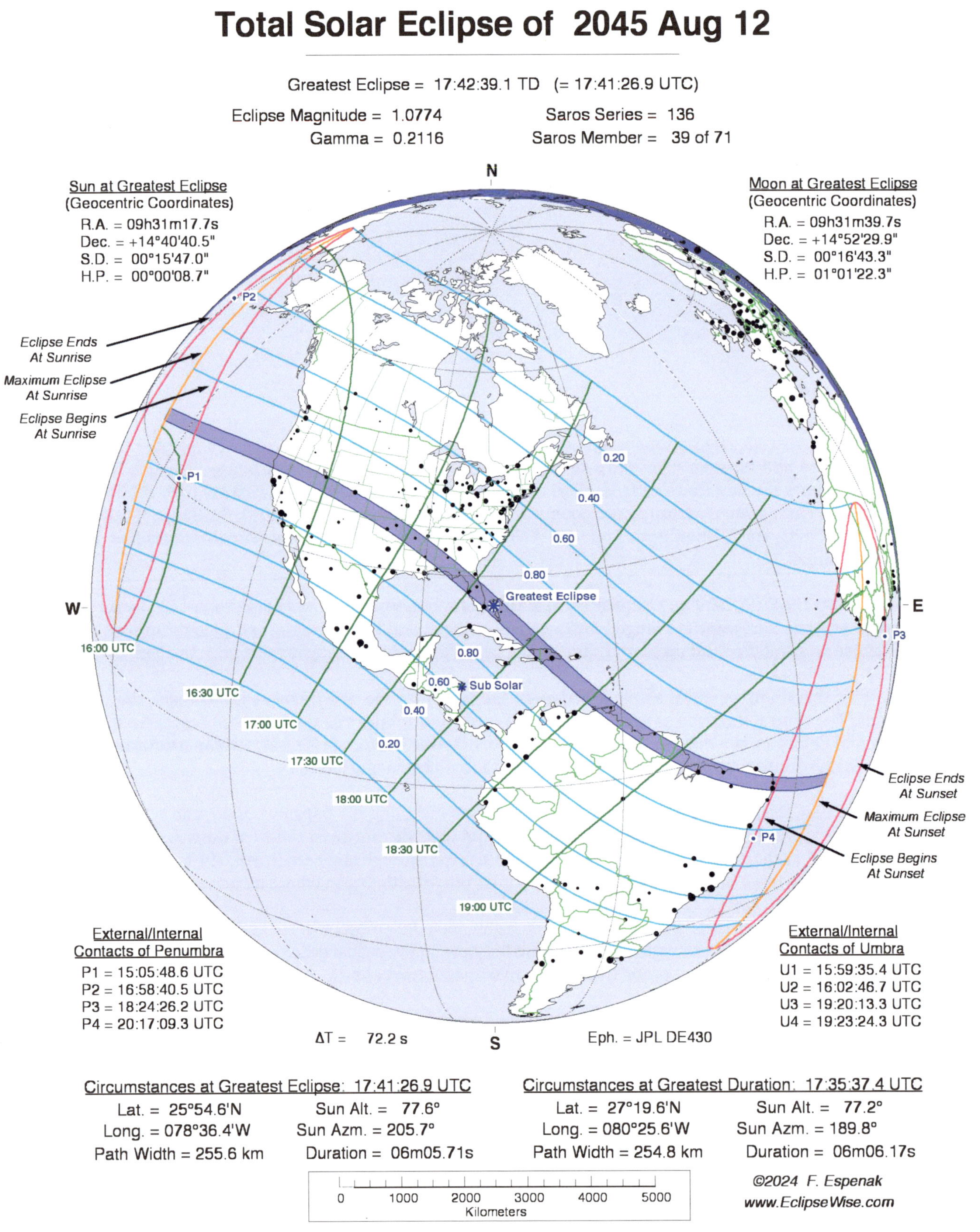

Figure 2 – Orthographic projection map of the Total Solar Eclipse of 2045.
*Based on map from **21st Century Canon of Solar Eclipses** by Fred Espenak (2017)*

2 – Orthographic Projection Map of the 2045 Total Solar Eclipse Path

The orthographic projection map of Earth in *Figure 2* shows the path of penumbral (partial) and umbral (total) shadows of the Moon during the eclipse.

The limits of the Moon's penumbral shadow define the region of visibility of the partial eclipse. It covers about half of Earth's daylight hemisphere and consists of several distinct zones or limits. At the northern and southern boundaries lie the limits of the penumbra's path. Great loops at the western and eastern extremes of the penumbra's path identify the areas where the eclipse begins and ends at sunrise and sunset, respectively.

Bisecting the *eclipse begins and ends at sunrise and sunset* loops is the curve of maximum eclipse at sunrise (western loop) and sunset (eastern loop). Points *P1* and *P4* mark the coordinates where the penumbral shadow first contacts (partial eclipse begins) and last contacts (partial eclipse ends) Earth's surface. The path of the umbral shadow (*path of totality*) appears within the penumbral path. Curves of constant eclipse magnitude define points where the maximum magnitude has values of 0.2, 0.4, 0.6, and 0.8.

Above the map is the instant of greatest eclipse[2], in Terrestrial Time[3] (TT) and Coordinated Universal Time[4] (UTC). The *Eclipse Magnitude* is the geocentric ratio of diameters of the Moon and the Sun. *Gamma* is the minimum distance of the Moon's shadow axis from Earth's center in Earth radii at greatest eclipse. Finally, the *Saros Series* of the eclipse is listed, followed by the *Saros Member* that gives the sequence position of the eclipse in the Saros, and the total number of eclipses in the series

In the upper left and right corners are the geocentric coordinates of the Sun and the Moon, respectively, at the instant of greatest eclipse. They are:

- R.A. – Right Ascension
- Dec. – Declination
- S.D. – Apparent Semi-Diameter
- H.P. – Horizontal Parallax

To the lower left are the contact times of the Moon's penumbral shadow with Earth. They mark the times when the partial eclipse begins and ends, globally. To the lower right are the contact times of the Moon's umbral shadow with Earth. They mark the times when the total eclipse path begins and ends, globally.

At bottom left are the geographic coordinates of the position of greatest eclipse along with the local circumstances at that location (i.e., Sun altitude, Sun azimuth, path width and duration of totality).

At bottom right are the corresponding values for the position of greatest duration[5]. Although the geographic locations of greatest duration and greatest eclipse differ by about 240 km (149 miles) along the eclipse track, the difference in the duration of totality is only 0.46 second.

Photo 2 – Time sequence of the Total Solar Eclipse of August 21, 2017 from Casper, Wyoming.
This sequence shows the partial phases, the diamond ring effect, and the solar corona during totality.
© 2017 F. Espenak, www.MrEclipse.com

---

[2] The instant of greatest eclipse occurs when the Moon's shadow axis passes closest to Earth's center.

[3] Terrestrial Time or TT is a time scale established by astronomers to calculate the location or planetary position of other objects as they relate to the center of Earth and its rotation.

[4] Coordinated Universal Time or UTC is the primary time standard that regulates world clocks and time. It is based on International Atomic Time (TAI) with leap seconds added at irregular intervals to compensate for the slowing of Earth's rotation. UTC is not adjusted for daylight saving time. It is effectively a successor to Greenwich Mean Time (GMT).

[5] Greatest duration is the time (and location) when the length of totality reaches its maximum value.

Figure 3 – Path of the Total Solar Eclipse of Aug 12, 2045 through the USA.

3 – Path of Totality

The path of the Moon's umbral shadow begins at 16:01 UTC in the North Pacific about 1590 km (990 mi) north of the Hawaiian Islands. Along the sunrise terminator, the duration of totality is 3m 1s (i.e., 3 minutes 1 seconds) as seen from the center of the narrow, 204 km (127 mi) wide path. No landfall occurs during the first 15 minutes of the central eclipse as the shadow sweeps 3140 km (1950 mi) northeast across the Pacific.

The path of the total eclipse across the USA begins at 16:16 UTC (9:16 am PDT) along the Pacific coast of northern California. The central duration of totality is 4m 23s, the Sun's altitude is 31° and the track width is 228 km (142 mi). The largest California city in the path is Redding (pop. 93,000) where its citizens will witness 4m 25s of totality.

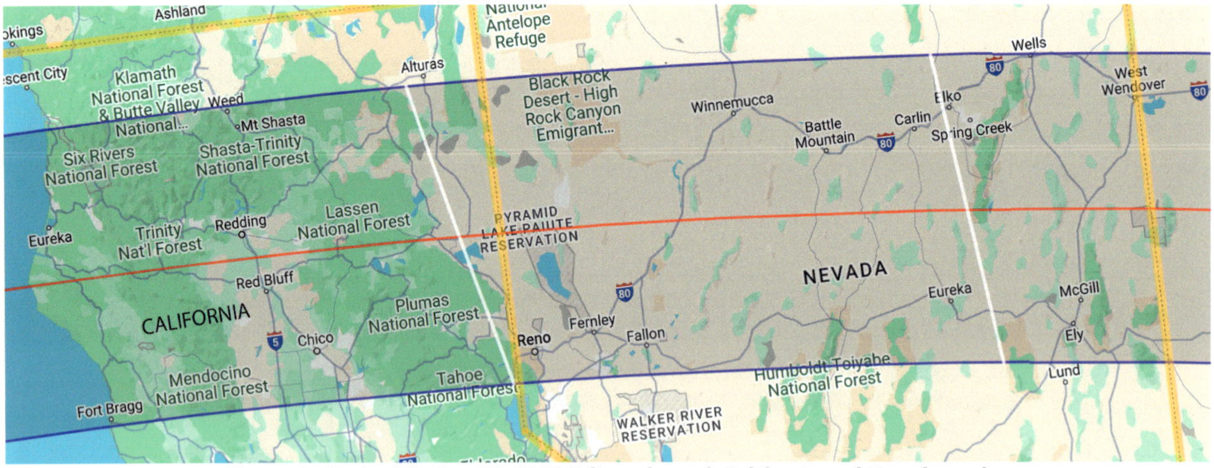

Figure 4 – Path of the Aug 12, 2045 Eclipse through California and Nevada, and.

Continuing over the Sierra Nevada Range, the track enters Nevada. Lady luck shines of Reno (pop. 243,000) which lies within the path. Although 93 km (58 mi.) south of the central line, Reno's population still enjoys 2m 45s of totality. Las Vegas, Carson City, and Lake Tahoe do not fare as well because they all lie outside the path and only witness a deep partial eclipse.

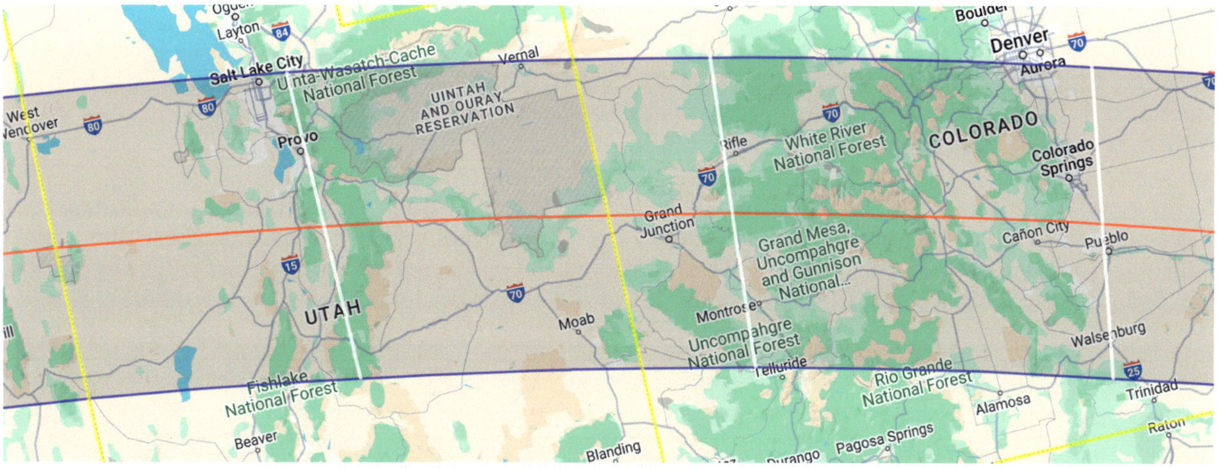

Figure 5 – Path of the Aug 12, 2045 Eclipse through Utah and Colorado.

Racing across Utah, the state's two largest cities Salt Lake City and Provo lie within the umbra's path. As it exits the "Beehive State," the path begins to curve to the southeast and the central duration exceeds five minutes. Colorado Springs is the next large city (pop. 486,000) in the path with a duration of 5m 4s and a Sun altitude of 50°. Denver, Colorado's capital, is just outside the northern path limit although some southern regions within the metropolitan area will witness a brief totality.

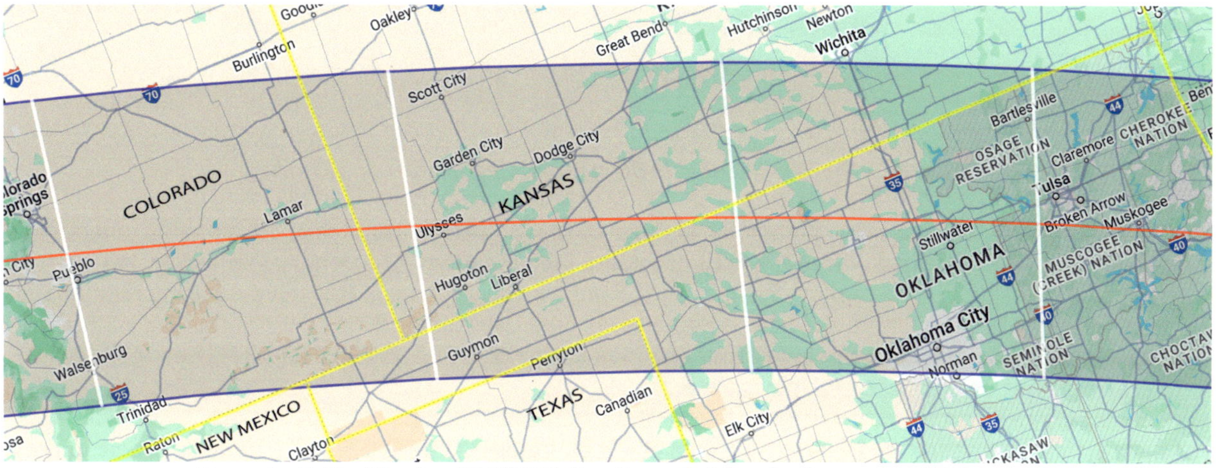

Figure 6 – Path of the Aug 12, 2045 Eclipse through Colorado, Kansas, and Oklahoma.

A tiny corner (approximately 40 sq. km) of northeastern New Mexico is grazed by the southern edge of the path as it enters the panhandle of Oklahoma. The path also includes southwestern Kansas and northeastern corner of Texas' panhandle. In Oklahoma, Oklahoma City (pop. 695,000) is the largest city in the path with a duration of 3m 26s, but smaller Tulsa (pop. 412,000) is closer to center where totality lasts 5m 33s.

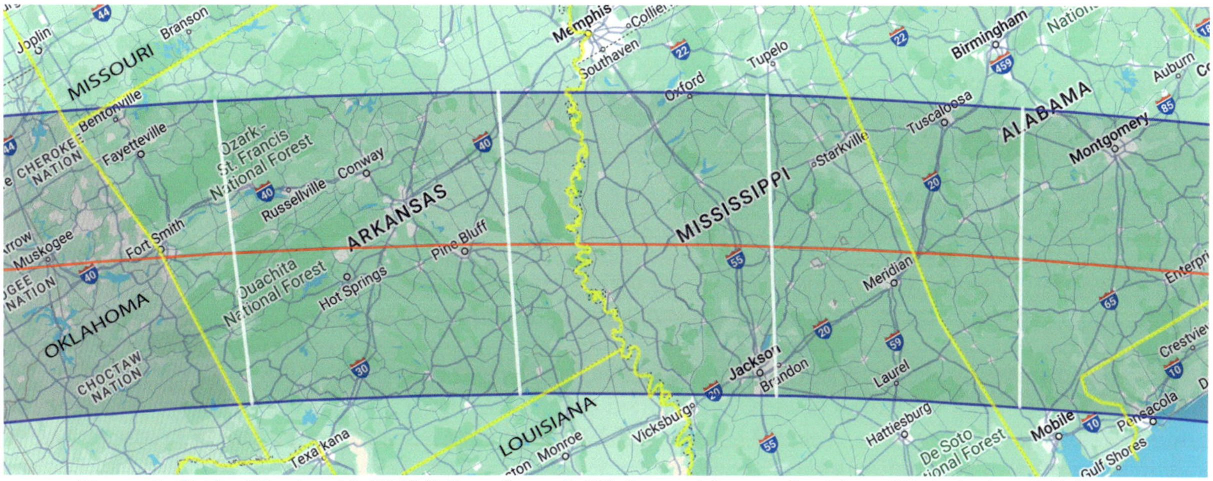

Figure 7 – Path of the Aug 12, 2045 Eclipse through Oklahoma, Arkansas, Louisiana, Mississippi, and Alabama.

The southwestern corner (approximately 230 sq. km) of Missouri is clipped by the path as it crosses into Arkansas. Just over the border, the inhabitants of Fort Smith (pop. 90,000) get totality lasting 5m 44s with the Sun 62° high. But the largest city in the Arkansas path, also its state capital, is Little Rock (pop. 203,000) which witnesses a total eclipse lasting 5m 36s.

As the umbra's track traverses the Mississippi River, it enters the northeast corner of Louisiana and cuts across Mississippi where the state capital, Jackson (pop. 146,000), receives 3m 10s of totality. Proceeding on to Alabama, Montgomery (pop. 197,000), another state capital, lies within the path with a total eclipse lasting 3m 40s. Spanning the "Cotton State," the track crosses the southwestern corner of Georgia before entering Florida.

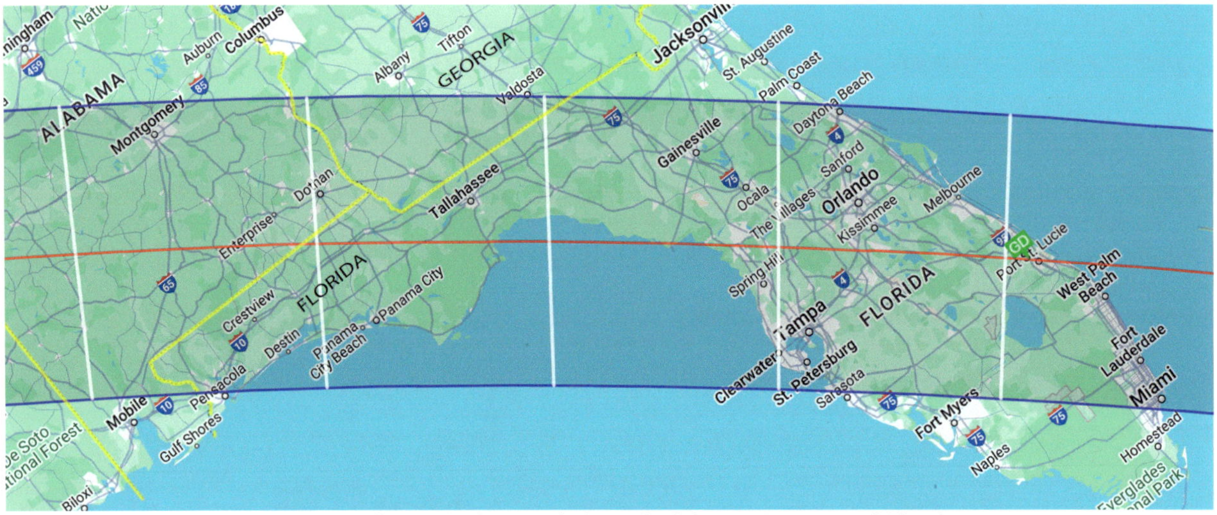

Figure 8 – The 2045 eclipse path through Alabama, Georgia, and Florida.

The "Sunshine State" is the most populated section of the eclipse track. Sixteen cities with populations exceeding 100,000 are located within the path of totality including Fort Lauderdale, Gainesville, Orlando, St. Petersburg, Tallahassee (state capital), Tampa, West Palm Beach, and Miami. It is not surprising that most of Florida's largest cities are within the path because it encompasses about 80% of the state. The eclipse also reaches its maximum with the point of *Greatest Duration* occurring at 17:36:27 UTC (1:36:27 pm EDT) in Port St. Lucie. The duration on the central line is 6m 6.2s.

The path of totality through the USA crosses parts of fifteen states (California, Nevada, Utah, Colorado, Kansas, Texas, New Mexico, Oklahoma, Missouri, Arkansas, Mississippi, Louisiana, Alabama, Georgia, and Florida). It is 4360 km (2710 mi) long and the Moon's umbral shadow traverses it in 81 minutes – an average velocity of 3230 km/hr (2007 mi/hr).

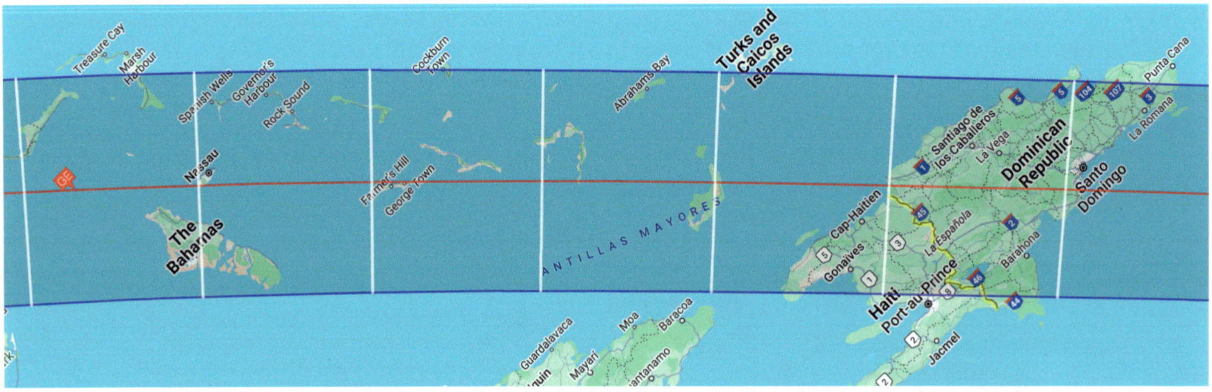

Figure 9 – The 2045 eclipse path through the Bahamas, Turks and Caicos, and Hispaniola.

After leaving the Atlantic coast of Florida, the shadow sweeps over the Lucayan Archipelago consisting of the Bahamas and the Turks and Caicos Islands. The point of *Greatest Eclipse* (the shadow axis passes closest to the center of Earth) occurs at sea at 17:41:27 UTC (1:41:27 EDT) about 66 km (41 mi) south of Freeport, Grand Bahama. The duration of totality is 6m 5.7s and the Sun's altitude is 78°.

Hispaniola is the most populous island in the West Indies. Its two nations (Haiti and the Dominican Republic) are both within the path of totality. Santo Domingo, the capital of the Dominican Republic, is one of the Caribbean's oldest cities. Its 4,275,000 residents will experience a total eclipse lasting 5m 44s at 18:10 UTC (2:10 pm AST).

Crossing the eastern Caribbean Sea, Isla La Blanquilla and Isla de Margarita are in the path just before the shadow reaches the northern coastline of Venezuela. While the entire island of Trinidad is inside the path, the island of Tobago is outside.

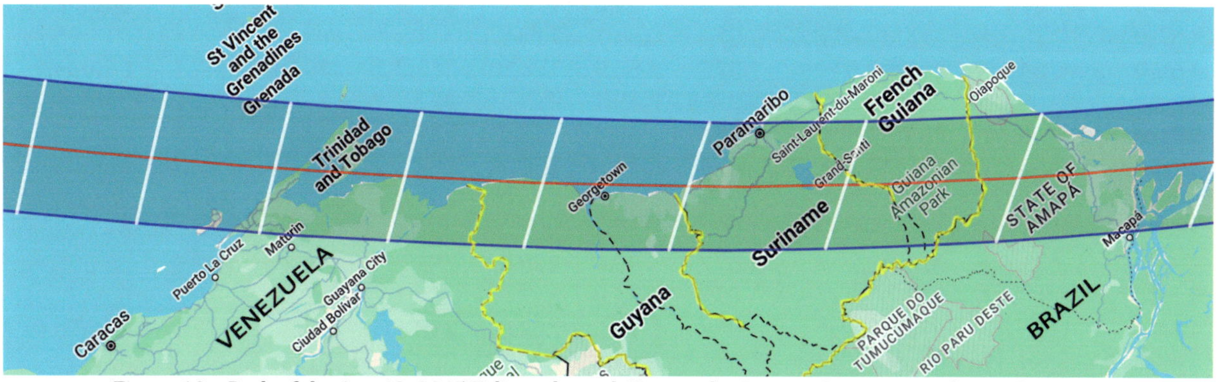

Figure 10 – Path of the Aug 12, 2045 Eclipse through Venezuela, Guyana, Suriname, and French Guiana.

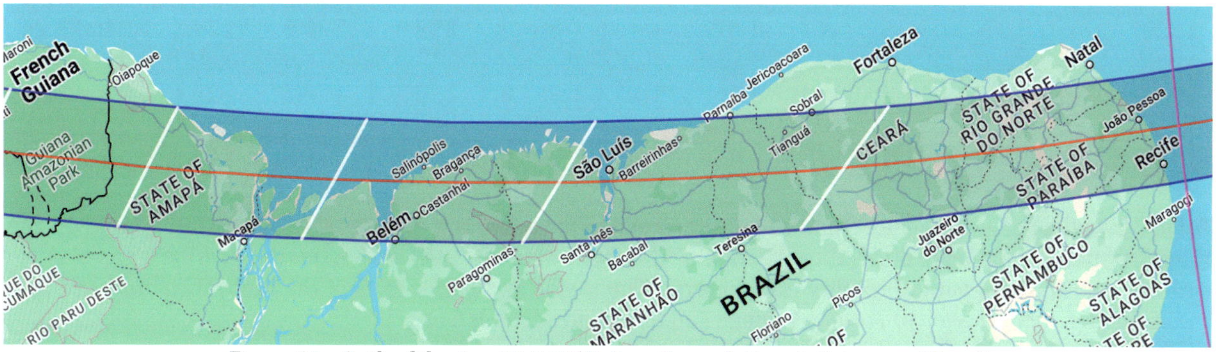

Figure 11 – Path of the Aug 12, 2045 Eclipse through French Guiana and Brazil.

The eclipse track remains close to the Atlantic coast of South America as it crosses Guyana, Suriname, French Guiana, and Brazil. São Luís is the capital and largest city of the Brazilian state of Maranhão. Its population of 1,536,000 residents will enjoy a totality lasting 3m 59s with the late afternoon Sun 26° above the horizon.

Leaving Brazil at 19:19 UTC (4:18 pm BRT), the umbra races 1460 km (905 mi) across the Atlantic before reaching the end of the track at 19:22 UTC and returning to space. Over a period of 3 hours 21 minutes, the umbra travels along a 15,199 km (9444 mi) track that covers an area equal to 0.71% of Earth's surface.

Table 1 – Time Zones used by the USA and Countries within the 2045 Path of Totality

Time Zone Name	Abbrev.	UTC Offset	Regions in 2045 Path
Pacific Daylight Time	PDT	UTC - 7 hrs	CA, NV
Mountain Daylight Time	MDT	UTC - 6 hrs	UT, CO, NM
Central Daylight Time	CDT	UTC - 5 hrs	KS, OK, TX, AR, LA, MS, AL, west FL
Eastern Daylight Time	EDT	UTC - 4 hrs	GE, FL, Bahamas, Hati
Atlantic Standard Time	AST	UTC - 4 hrs	Dominican Republic, Trinidad
Venezuela Standard Time	VST	UTC - 4 hrs	Venezuela
Guyana Time	GYT	UTC - 4 hrs	Guyana
Suriname Time	SRT	UTC - 3 hrs	Suriname
Brasilia Time	BRT	UTC - 3 hrs	Brazil

Table 2 – The 2045 Path of Totality

UTC	Latitude	Longitude	Diameter Ratio	Sun Alt	Sun Azm	Path Width km	Major Axis km	Minor Axis km	Duration of Totality
16:01.2	35°07.7'N	159°41.0'W	1.059	0.0	71.9	204.2	—	193.5	03m01.5s
16:10	40°02.6'N	132°35.7'W	1.066	23.6	90.3	223.8	542.4	216.7	04m01.7s
16:20	40°22.8'N	120°26.1'W	1.070	34.7	100.6	231.6	398.3	226.6	04m34.0s
16:30	39°41.7'N	111°38.9'W	1.072	43.2	109.1	237.1	340.8	233.3	04m58.8s
16:40	38°28.4'N	104°39.4'W	1.073	50.5	117.0	241.5	309.0	238.3	05m18.8s
16:50	36°54.4'N	098°50.8'W	1.075	56.9	124.9	245.0	289.0	242.1	05m34.9s
17:00	35°05.5'N	093°53.2'W	1.076	62.7	133.4	247.9	275.7	245.0	05m47.4s
17:10	33°05.4'N	089°33.8'W	1.076	68.0	143.6	250.3	266.7	247.2	05m56.6s
17:20	30°56.3'N	085°43.5'W	1.077	72.5	157.0	252.3	260.8	248.8	06m02.7s
17:30	28°39.6'N	082°15.3'W	1.077	76.0	176.0	254.0	257.4	249.7	06m05.7s
17:40	26°15.9'N	079°03.2'W	1.077	77.6	201.7	255.4	256.1	250.1	06m05.9s
17:50	23°45.8'N	076°02.3'W	1.077	76.8	228.5	256.7	256.7	249.9	06m03.4s
18:00	21°09.2'N	073°07.9'W	1.077	73.8	249.0	257.7	259.5	249.1	05m58.2s
18:10	18°25.8'N	070°15.1'W	1.077	69.5	262.6	258.4	264.5	247.7	05m50.6s
18:20	15°34.8'N	067°18.9'W	1.076	64.5	271.5	258.7	272.4	245.7	05m40.5s
18:30	12°35.0'N	064°12.9'W	1.075	58.8	277.6	258.3	284.2	243.0	05m28.1s
18:40	09°24.2'N	060°48.5'W	1.074	52.6	281.9	256.9	301.7	239.5	05m13.0s
18:50	05°59.0'N	056°52.8'W	1.072	45.6	284.9	253.9	328.9	234.8	04m55.2s
19:00	02°12.6'N	052°02.0'W	1.070	37.5	286.8	248.4	375.9	228.7	04m33.8s
19:10	02°10.8'S	045°20.3'W	1.067	27.4	287.8	238.7	478.2	220.1	04m07.0s
19:20	08°26.8'S	032°11.3'W	1.062	10.7	286.8	218.0	1101.4	204.1	03m25.1s
19:21.8	11°35.8'S	022°17.7'W	1.059	0.0	285.0	203.7	—	193.3	03m01.3s

Table 2 lists the geographic coordinates of the central line for every ten minutes UTC. *Diameter Ratio* gives the Moon's apparent diameter divided by the Sun's. The Sun's altitude (*Alt*) and azimuth (*Azm*) are given along with the width of the path of totality (kilometers). The *major* and *minor* axes of the umbral shadow are also listed. Finally, the *Duration of Totality* is given (minutes and seconds).

4 – Local Circumstances for the Eclipse

Local circumstances for the ***Total Eclipse*** for 162 cities are found in Table 3 (USA – Alabama to Kansas), Table 4 (USA - Florida), Table 5 (USA – Mississippi to Utah), and Table 6 (West Indies and South America). All times are given in local time (see Table 1 for Time Zones). The *Duration of Totality* is the period between the beginning and end of totality. These times include the effects of the Moon's limb profile. The Sun's altitude is given at the instant of mid-totality.

Local circumstances for the ***Partial Eclipse*** for 102 cities are found in Table 7 (USA) and Table 8 (North, Central, and South America). All times are given in local time and include *Daylight Saving Time*. The Sun's altitude is given at the instant of maximum eclipse. *Eclipse Magnitude* is given as the fraction of the Sun's *diameter* covered by the Moon at maximum eclipse.

The 2045 Total Solar Eclipse Circumstances Calculator is an interactive web page that can quickly calculate the local circumstances for the solar eclipse from any geographic location not included in Tables 3 through 8. The URL address for the calculator is:

https://www.eclipsewise.com/solar/SEcirc/2001-2100/SE2045Aug12Tcirc.html

Table 3 – Total Eclipse Circumstances for Alabama to Kansas

Location	Partial Eclipse Begins	Total Eclipse Begins	Total Eclipse Ends	Partial Eclipse Ends	Duration of Totality	Sun Altitude
Alabama						
Dothan	10:59:06 am	12:17:21 pm	12:23:05 pm	01:40:23 pm	05m40s	72°
Enterprise	10:58:12 am	12:16:16 pm	12:22:11 pm	01:39:29 pm	05m53s	72°
Montgomery	10:56:29 am	12:15:03 pm	12:18:43 pm	01:36:51 pm	03m40s	71°
Northport	10:53:32 am	12:11:59 pm	12:14:46 pm	01:33:14 pm	02m47s	69°
Prattville	10:56:07 am	12:14:37 pm	12:18:20 pm	01:36:26 pm	03m44s	71°
Selma	10:55:13 am	12:13:00 pm	12:18:05 pm	01:35:40 pm	05m06s	70°
Tuscaloosa	10:53:41 am	12:12:07 pm	12:15:00 pm	01:33:25 pm	02m54s	69°
Arkansas						
Benton	10:44:35 am	11:59:56 am	12:05:45 pm	01:22:46 pm	05m50s	64°
Conway	10:44:35 am	12:00:03 pm	12:05:08 pm	01:22:16 pm	05m05s	64°
Fayetteville	10:41:35 am	11:56:34 am	12:00:37 pm	01:17:52 pm	04m09s	62°
Florence	10:46:29 am	12:02:30 pm	12:08:09 pm	01:25:36 pm	05m41s	66°
Fort Smith	10:41:25 am	11:55:53 am	12:01:38 pm	01:18:21 pm	05m44s	62°
Hot Springs	10:43:53 am	11:59:12 am	12:04:52 pm	01:22:03 pm	05m44s	64°
Jacksonville	10:45:11 am	12:00:43 pm	12:06:03 pm	01:23:08 pm	05m16s	64°
Japton	10:42:10 am	11:57:24 am	12:01:15 pm	01:18:38 pm	03m56s	62°
Little Rock	10:44:58 am	12:00:22 pm	12:06:01 pm	01:23:01 pm	05m36s	64°
Pine Bluff	10:45:39 am	12:01:19 pm	12:07:08 pm	01:24:17 pm	05m51s	65°
Rogers	10:41:35 am	11:57:20 am	11:59:38 am	01:17:34 pm	02m14s	62°
Russellville	10:43:25 am	11:58:29 am	12:03:51 pm	01:20:46 pm	05m18s	63°
Springdale	10:41:36 am	11:56:51 am	12:00:17 pm	01:17:45 pm	03m29s	62°
California						
Chico	08:13:52 am	09:16:40 am	09:20:05 am	10:28:46 am	03m25s	33°
Eureka	08:13:17 am	09:14:36 am	09:18:44 am	10:25:41 am	04m07s	31°
Paradise	08:14:00 am	09:16:49 am	09:20:23 am	10:29:04 am	03m34s	34°
Red Bluff	08:13:52 am	09:15:59 am	09:20:18 am	10:28:12 am	04m22s	33°
Redding	08:14:01 am	09:15:56 am	09:20:21 am	10:28:00 am	04m25s	33°
Susanville	08:14:52 am	09:17:30 am	09:22:02 am	10:30:25 am	04m33s	35°
Truckee	08:14:37 am	-	09:20:10 am	10:31:07 am	00m00s	35°
Colorado						
Canon City	09:26:31 am	10:35:54 am	10:40:59 am	11:55:06 am	05m09s	49°
Colorado Springs	09:27:30 am	10:37:04 am	10:42:11 am	11:56:16 am	05m04s	50°
Englewood	09:27:31 am	10:38:35 am	10:39:45 am	11:55:19 am	01m04s	50°
Grand Junction	09:23:40 am	10:31:39 am	10:36:40 am	11:49:45 am	05m04s	46°
Littleton	09:27:29 am	10:38:18 am	10:40:02 am	11:55:19 am	01m41s	50°
Parker	09:27:43 am	10:38:23 am	10:40:40 am	11:55:47 am	02m16s	50°
Pueblo	09:27:40 am	10:37:32 am	10:42:41 am	11:57:09 am	05m13s	51°
Georgia						
Bainbridge	12:00:53 pm	01:19:30 pm	01:24:46 pm	02:42:14 pm	05m16s	73°
Thomasville	12:02:04 pm	01:21:15 pm	01:25:29 pm	02:43:19 pm	04m19s	73°
Kansas						
Derby	10:36:51 am	11:51:31 am	11:52:55 am	01:10:35 pm	01m19s	58°
Dodge City	10:33:10 am	11:45:03 am	11:50:06 am	01:05:37 pm	05m01s	55°
Garden City	10:32:05 am	11:43:32 am	11:48:37 am	01:03:53 pm	05m03s	54°
Liberal	10:32:04 am	11:43:56 am	11:48:58 am	01:04:47 pm	05m01s	55°

Table 4 – Total Eclipse Circumstances for Florida

Location	Partial Eclipse Begins	Total Eclipse Begins	Total Eclipse Ends	Partial Eclipse Ends	Duration of Totality	Sun Altitude
Boca Raton	12:14:47 pm	01:35:13 pm	01:40:34 pm	02:57:39 pm	05m23s	78°
Boynton Beach	12:14:34 pm	01:34:44 pm	01:40:27 pm	02:57:20 pm	05m45s	78°
Bradenton	12:08:27 pm	01:30:16 pm	01:32:22 pm	02:51:39 pm	02m04s	77°
Clearwater	12:07:25 pm	01:28:00 pm	01:32:08 pm	02:50:26 pm	04m09s	77°
Coral Gables	12:15:17 pm	01:37:47 pm	01:39:33 pm	02:58:29 pm	01m39s	79°
Coral Springs	12:14:40 pm	01:35:15 pm	01:40:21 pm	02:57:36 pm	05m05s	78°
Davie	12:14:53 pm	01:36:04 pm	01:40:12 pm	02:57:57 pm	04m09s	78°
Daytona Beach	12:09:24 pm	01:30:40 pm	01:32:09 pm	02:50:49 pm	01m22s	75°
De Land	12:09:03 pm	01:29:03 pm	01:33:14 pm	02:50:46 pm	04m15s	76°
Deerfield Beach	12:14:48 pm	01:35:18 pm	01:40:33 pm	02:57:41 pm	05m17s	78°
Delray Beach	12:14:40 pm	01:34:54 pm	01:40:30 pm	02:57:27 pm	05m38s	78°
Deltona	12:09:20 pm	01:29:12 pm	01:33:43 pm	02:51:07 pm	04m35s	76°
Fort Lauderdale	12:15:01 pm	01:35:51 pm	01:40:35 pm	02:58:01 pm	04m39s	78°
Fort Pierce	12:12:50 pm	01:32:34 pm	01:38:36 pm	02:55:12 pm	06m03s	77°
Fort Walton Beach	10:57:40 am	12:17:05 pm	12:21:01 pm	01:39:48 pm	03m57s	73°
Gainesville	12:06:22 pm	01:25:56 pm	01:30:28 pm	02:48:01 pm	04m36s	75°
Hialeah	12:15:07 pm	01:37:15 pm	01:39:41 pm	02:58:19 pm	02m27s	78°
Hollywood	12:15:08 pm	01:36:17 pm	01:40:28 pm	02:58:10 pm	04m12s	78°
Kissimmee	12:09:42 pm	01:29:11 pm	01:35:11 pm	02:52:00 pm	06m00s	76°
Lakeland	12:08:56 pm	01:28:34 pm	01:34:26 pm	02:51:37 pm	05m53s	77°
Largo	12:07:30 pm	01:28:13 pm	01:32:09 pm	02:50:33 pm	03m58s	77°
Lauderhill	12:14:50 pm	01:35:47 pm	01:40:18 pm	02:57:51 pm	04m28s	78°
Margate	12:14:42 pm	01:35:23 pm	01:40:21 pm	02:57:40 pm	04m54s	78°
Melbourne	12:11:30 pm	01:31:11 pm	01:36:46 pm	02:53:35 pm	05m33s	77°
Miami	12:15:24 pm	01:37:26 pm	01:40:04 pm	02:58:33 pm	02m42s	78°
Miami Beach	12:15:29 pm	01:37:24 pm	01:40:18 pm	02:58:38 pm	03m00s	78°
Miramar	12:14:59 pm	01:36:28 pm	01:40:03 pm	02:58:05 pm	03m41s	78°
North Miami	12:15:14 pm	01:36:51 pm	01:40:14 pm	02:58:20 pm	03m30s	78°
Ocala	12:07:16 pm	01:26:35 pm	01:32:08 pm	02:49:14 pm	05m31s	75°
Orlando	12:09:26 pm	01:28:55 pm	01:34:40 pm	02:51:33 pm	05m42s	76°
Palm Bay	12:11:37 pm	01:31:18 pm	01:36:56 pm	02:53:43 pm	05m36s	77°
Panama City	10:59:36 am	12:18:37 pm	12:23:46 pm	01:41:48 pm	05m10s	73°
Pembroke Pines	12:14:59 pm	01:36:22 pm	01:40:09 pm	02:58:05 pm	03m50s	78°
Pinellas Park	12:07:44 pm	01:28:35 pm	01:32:18 pm	02:50:48 pm	03m47s	77°
Plantation	12:14:46 pm	01:35:47 pm	01:40:11 pm	02:57:48 pm	04m22s	78°
Pompano Beach	12:14:51 pm	01:35:30 pm	01:40:32 pm	02:57:48 pm	05m00s	78°
Port Orange	12:09:32 pm	01:30:32 pm	01:32:35 pm	02:51:00 pm	01m58s	76°
Port St. Lucie	12:12:57 pm	01:32:43 pm	01:38:48 pm	02:55:24 pm	06m06s	77°
St. Petersburg	12:07:54 pm	01:28:49 pm	01:32:27 pm	02:50:59 pm	03m42s	77°
Sunrise	12:14:56 pm	01:36:00 pm	01:40:20 pm	02:57:58 pm	04m19s	78°
Tallahassee	12:01:53 pm	01:20:29 pm	01:26:20 pm	02:43:33 pm	05m47s	74°
Tamarac	12:14:48 pm	01:35:36 pm	01:40:22 pm	02:57:47 pm	04m42s	78°
Tampa	12:08:06 pm	01:28:14 pm	01:33:16 pm	02:51:01 pm	05m00s	77°
Titusville	12:10:28 pm	01:30:28 pm	01:34:55 pm	02:52:16 pm	04m31s	76°
West Palm Beach	12:14:21 pm	01:34:21 pm	01:40:18 pm	02:57:01 pm	05m58s	78°
Winter Haven	12:09:24 pm	01:28:59 pm	01:34:59 pm	02:52:01 pm	06m00s	77°

Table 5 – Total Eclipse Circumstances for Mississippi to Utah

Location	Partial Eclipse Begins	Total Eclipse Begins	Total Eclipse Ends	Partial Eclipse Ends	Duration of Totality	Sun Altitude
Mississippi						
Brandon	10:50:12 am	12:08:22 pm	12:12:00 pm	01:30:52 pm	03m41s	68°
Clarksdale	10:47:59 am	12:04:13 pm	12:09:28 pm	01:26:46 pm	05m13s	66°
Cleveland	10:48:00 am	12:04:11 pm	12:10:03 pm	01:27:15 pm	05m53s	67°
Clinton	10:49:34 am	12:08:06 pm	12:10:53 pm	01:30:08 pm	02m54s	68°
Columbus	10:52:01 am	12:09:41 pm	12:13:30 pm	01:31:31 pm	03m52s	68°
Greenville	10:47:38 am	12:04:02 pm	12:09:34 pm	01:27:11 pm	05m34s	67°
Greenwood	10:49:02 am	12:05:25 pm	12:11:18 pm	01:28:32 pm	05m54s	67°
Jackson	10:49:49 am	12:08:14 pm	12:11:19 pm	01:30:25 pm	03m10s	68°
Laurel	10:52:06 am	12:11:15 pm	12:13:54 pm	01:33:18 pm	02m45s	70°
Meridian	10:52:19 am	12:09:32 pm	12:15:22 pm	01:32:56 pm	05m51s	69°
Pearl	10:49:55 am	12:08:24 pm	12:11:25 pm	01:30:34 pm	03m07s	68°
Ridgeland	10:49:50 am	12:07:35 pm	12:11:46 pm	01:30:20 pm	04m08s	68°
Starkville	10:51:21 am	12:08:30 pm	12:13:18 pm	01:30:55 pm	04m51s	68°
Nevada						
Elko	08:18:14 am	09:23:21 am	09:26:51 am	10:37:27 am	03m33s	39°
Ely	08:18:12 am	09:24:15 am	09:27:29 am	10:39:14 am	03m12s	40°
Fallon	08:15:33 am	09:20:02 am	09:22:50 am	10:33:11 am	02m44s	36°
Reno	08:14:55 am	09:18:57 am	09:21:45 am	10:31:40 am	02m45s	35°
Sparks	08:14:58 am	09:18:59 am	09:21:52 am	10:31:45 am	02m51s	35°
Winnemucca	08:16:58 am	09:21:14 am	09:24:40 am	10:34:31 am	03m28s	37°
Oklahoma						
Bartlesville	10:38:45 am	11:52:53 am	11:56:56 am	01:13:49 pm	04m08s	60°
Broken Arrow	10:39:11 am	11:52:58 am	11:58:38 am	01:15:03 pm	05m37s	60°
Del City	10:37:02 am	11:51:52 am	11:54:59 am	01:12:54 pm	03m09s	59°
Edmond	10:36:54 am	11:51:02 am	11:55:16 am	01:12:33 pm	04m13s	59°
Enid	10:36:09 am	11:49:11 am	11:54:41 am	01:10:56 pm	05m34s	58°
Midwest City	10:37:04 am	11:51:48 am	11:55:07 am	01:12:57 pm	03m20s	59°
Moore	10:36:58 am	11:52:22 am	11:54:32 am	01:12:53 pm	02m12s	59°
Muskogee	10:39:54 am	11:53:57 am	11:59:39 am	01:16:13 pm	05m43s	61°
Oklahoma City	10:36:54 am	11:51:32 am	11:54:57 am	01:12:42 pm	03m26s	59°
Ponca City	10:37:13 am	11:50:24 am	11:55:48 am	01:11:58 pm	05m21s	59°
Shawnee	10:37:47 am	11:52:31 am	11:56:13 am	01:13:58 pm	03m42s	60°
Stillwater	10:37:21 am	11:50:46 am	11:56:18 am	01:12:44 pm	05m36s	59°
Tulsa	10:38:59 am	11:52:42 am	11:58:18 am	01:14:42 pm	05m33s	60°
Utah						
Moab	09:22:37 am	10:31:06 am	10:34:33 am	11:48:21 am	03m25s	46°
Murray	09:21:12 am	10:28:30 am	10:30:55 am	11:43:28 am	02m26s	43°
Orem	09:21:12 am	10:27:55 am	10:31:59 am	11:43:55 am	04m05s	43°
Provo	09:21:11 am	10:27:50 am	10:32:10 am	11:44:01 am	04m18s	43°
Salt Lake City	09:21:13 am	10:28:55 am	10:30:30 am	11:43:24 am	01m32s	43°
Sandy City	09:21:25 am	10:28:45 am	10:31:22 am	11:43:56 am	02m40s	43°
South Jordan	09:20:59 am	10:27:45 am	10:31:13 am	11:43:17 am	03m32s	43°
Taylorsville	09:21:09 am	10:28:25 am	10:30:53 am	11:43:23 am	02m28s	43°
West Jordan	09:21:05 am	10:28:08 am	10:31:01 am	11:43:20 am	02m58s	43°
West Valley	09:21:07 am	10:28:24 am	10:30:47 am	11:43:18 am	02m23s	43°

Table 6 – Total Eclipse Circumstances for West Indies and South America

Location	Partial Eclipse Begins	Total Eclipse Begins	Total Eclipse Ends	Partial Eclipse Ends	Duration of Totality	Sun Altitude
Bahamas						
Freeport	12:17:14 pm	01:37:13 pm	01:43:04 pm	02:59:19 pm	05m47s	77°
Nassau	12:22:03 pm	01:42:13 pm	01:48:14 pm	03:04:00 pm	06m01s	77°
Matthew Town	12:36:13 pm	01:56:44 pm	02:02:06 pm	03:16:46 pm	05m25s	74°
Brazil						
Anil	03:04:54 pm	04:08:58 pm	04:12:56 pm	05:10:24 pm	03m57s	26°
Belem	02:58:50 pm	04:06:33 pm	04:08:05 pm	05:08:40 pm	01m30s	31°
Cabo	03:18:44 pm	04:18:44 pm	04:20:15 pm	05:14:37 pm	01m35s	14°
Caico	03:15:34 pm	04:15:41 pm	04:19:18 pm	05:13:38 pm	03m39s	17°
Camarajibe	03:18:31 pm	04:17:51 pm	04:20:39 pm	05:14:27 pm	02m46s	14°
Campina Grande	03:17:10 pm	04:16:41 pm	04:20:07 pm	05:14:03 pm	03m29s	15°
Castanhal	02:59:15 pm	04:05:39 pm	04:09:14 pm	05:08:39 pm	03m35s	31°
Joao Pessoa	03:17:44 pm	04:16:49 pm	04:20:16 pm	05:13:50 pm	03m26s	14°
Recife	03:18:36 pm	04:17:54 pm	04:20:40 pm	05:14:27 pm	02m45s	14°
Sao Luis	03:04:51 pm	04:08:56 pm	04:12:54 pm	05:10:23 pm	03m57s	26°
Vitoria de Santo Antao	03:18:25 pm	04:18:20 pm	04:20:13 pm	05:14:33 pm	01m56s	14°
Dominican Republic						
Benemerita de San Cri.	12:48:00 pm	02:07:23 pm	02:13:10 pm	03:25:41 pm	05m49s	69°
Concepcion de la Vega	12:45:42 pm	02:05:15 pm	02:10:50 pm	03:23:44 pm	05m33s	70°
La Romana	12:50:18 pm	02:10:06 pm	02:13:59 pm	03:26:52 pm	03m54s	68°
San Francisco de Mac.	12:46:06 pm	02:05:48 pm	02:10:44 pm	03:23:51 pm	04m58s	70°
San Juan de la Maguana	12:45:00 pm	02:04:53 pm	02:10:23 pm	03:23:42 pm	05m32s	71°
San Pedro de Macoris	12:49:33 pm	02:09:02 pm	02:13:46 pm	03:26:27 pm	04m47s	68°
Santiago de los Cabal.	12:44:55 pm	02:04:32 pm	02:10:02 pm	03:23:05 pm	05m29s	70°
Santo Domingo	12:48:19 pm	02:07:39 pm	02:13:20 pm	03:25:48 pm	05m40s	69°
French Guiana						
Saul	02:44:58 pm	03:54:50 pm	03:59:10 pm	05:01:30 pm	04m18s	40°
Guyana						
Georgetown	02:31:54 pm	03:44:51 pm	03:49:46 pm	04:54:51 pm	04m56s	48°
Haiti						
Cap-Haitien	01:41:20 pm	03:01:25 pm	03:07:00 pm	04:20:48 pm	05m37s	72°
Gonaives	01:40:54 pm	03:02:14 pm	03:05:51 pm	04:20:47 pm	03m41s	73°
Suriname						
Paramaribo	02:38:21 pm	03:50:38 pm	03:53:12 pm	04:57:43 pm	02m33s	43°
Trinidad & Tobago						
Port of Spain	01:19:06 pm	02:34:28 pm	02:39:23 pm	03:46:52 pm	04m53s	54°
San Fernando	01:19:52 pm	02:35:05 pm	02:40:15 pm	03:47:29 pm	05m11s	54°
Turks & Caicos Is.						
Providenciales Island	11:37:39 am	12:58:43 pm	01:01:44 pm	02:16:49 pm	02m59s	73°
West Caicos Island	11:37:29 am	12:57:58 pm	01:02:17 pm	02:16:52 pm	04m22s	73°
Venezuela						
Carupano	01:15:59 pm	02:32:35 pm	02:36:52 pm	03:45:34 pm	04m15s	56°
Porlamar	01:14:19 pm	02:31:52 pm	02:35:04 pm	03:44:36 pm	03m17s	57°
Tucupita	01:21:09 pm	02:38:07 pm	02:40:02 pm	03:48:57 pm	01m55s	54°

Table 7 - Partial Eclipse Circumstances for the USA

State	City	Partial Eclipse Begins	Maximum Eclipse	Partial Eclipse Ends	Eclipse Magnitude	Sun Altitude
Alabama	Birmingham	10:54:50 am	12:14:34 pm	01:34:05 pm	0.984	69°
	Mobile	10:54:51 am	12:15:55 pm	01:36:50 pm	0.993	71°
Alaska	Anchorage	06:41:55 am	07:22:57 am	08:05:25 am	0.342	15°
Arizona	Phoenix	08:20:26 am	09:29:53 am	10:45:29 am	0.825	45°
	Tucson	08:21:58 am	09:31:53 am	10:47:53 am	0.789	46°
California	Los Angeles	08:15:04 am	09:21:14 am	10:33:52 am	0.826	38°
	Sacramento	08:13:33 am	09:18:23 am	10:29:12 am	0.974	34°
	San Diego	08:16:12 am	09:22:39 am	10:35:37 am	0.782	39°
	San Francisco	08:12:46 am	09:17:12 am	10:27:45 am	0.946	33°
	San Jose	08:12:56 am	09:17:37 am	10:28:28 am	0.932	34°
Colorado	Denver	09:27:30 am	10:39:10 am	11:55:13 am	0.998	50°
Connecticut	Hartford	12:19:45 pm	01:26:38 pm	02:31:38 pm	0.548	62°
District of Columbia	Washington	12:11:12 pm	01:24:02 pm	02:35:06 pm	0.684	66°
Florida	Jacksonville	12:07:00 pm	01:28:29 pm	02:47:45 pm	0.983	74°
	Sarasota	12:08:43 pm	01:31:37 pm	02:52:01 pm	0.998	77°
Georgia	Atlanta	11:59:02 am	01:18:36 pm	02:37:17 pm	0.939	70°
Illinois	Chicago	10:51:40 am	12:04:22 pm	01:17:40 pm	0.760	61°
Indiana	Indianapolis	11:53:57 am	01:08:38 pm	02:23:28 pm	0.800	63°
Iowa	Des Moines	10:42:23 am	11:55:28 am	01:10:28 pm	0.839	58°
Kansas	Wichita	10:36:44 am	11:52:01 am	01:10:16 pm	0.997	58°
Kentucky	Louisville	11:54:45 am	01:10:49 pm	02:26:51 pm	0.836	65°
Louisiana	Baton Rouge	10:49:44 am	12:10:06 pm	01:31:23 pm	0.937	69°
	New Orleans	10:52:04 am	12:12:56 pm	01:34:15 pm	0.940	71°
Maryland	Baltimore	12:11:52 pm	01:24:07 pm	02:34:35 pm	0.669	65°
Massachusetts	Boston	12:23:11 pm	01:28:06 pm	02:31:05 pm	0.510	61°
	Springfield	12:19:57 pm	01:26:21 pm	02:30:53 pm	0.539	62°
Michigan	Detroit	11:59:36 am	01:10:52 pm	02:21:50 pm	0.687	62°
Minnesota	Minneapolis	10:44:21 am	11:54:04 am	01:05:28 pm	0.734	55°
Missouri	Kansas City	10:40:36 am	11:55:35 am	01:12:48 pm	0.924	59°
	St. Louis	10:47:12 am	12:03:03 pm	01:20:01 pm	0.886	62°
Nebraska	Omaha	10:38:56 am	11:51:59 am	01:07:30 pm	0.873	56°
Nevada	Las Vegas	08:17:22 am	09:25:20 am	10:39:23 am	0.905	41°
New Jersey	Newark	12:16:42 pm	01:25:52 pm	02:33:09 pm	0.596	63°
New Mexico	Albuquerque	09:25:33 am	10:37:57 am	11:55:35 am	0.912	50°
New York	Buffalo	12:07:22 pm	01:16:25 pm	02:24:30 pm	0.615	62°
	New York	12:17:03 pm	01:26:08 pm	02:33:18 pm	0.593	63°
North Carolina	Charlotte	12:04:53 pm	01:22:46 pm	02:39:03 pm	0.842	69°
	Winston-Salem	12:05:37 pm	01:22:34 pm	02:37:55 pm	0.809	69°
Ohio	Cincinnati	11:56:51 am	01:11:56 pm	02:26:46 pm	0.795	65°
	Cleveland	12:01:55 pm	01:13:45 pm	02:24:57 pm	0.690	63°
	Columbus	11:59:31 am	01:13:24 pm	02:26:47 pm	0.750	64°
	Dayton	11:57:22 am	01:11:45 pm	02:25:51 pm	0.773	64°
Oregon	Portland	08:17:50 am	09:20:11 am	10:27:17 am	0.866	32°
Pennsylvania	Philadelphia	12:14:43 pm	01:25:26 pm	02:34:15 pm	0.630	64°
	Pittsburgh	12:05:03 pm	01:17:30 pm	02:28:54 pm	0.693	64°
Rhode Island	Providence	12:22:29 pm	01:28:22 pm	02:32:16 pm	0.527	62°
South Carolina	Columbia	12:05:11 pm	01:24:07 pm	02:41:18 pm	0.877	71°
Tennessee	Memphis	10:48:24 am	12:06:49 pm	01:26:09 pm	0.986	66°
	Nashville	10:53:32 am	12:11:25 pm	01:29:21 pm	0.909	67°
Texas	Austin	10:39:39 am	11:57:14 am	01:17:58 pm	0.842	62°
	Dallas	10:39:08 am	11:56:43 am	01:17:13 pm	0.931	62°
	El Paso	09:27:05 am	10:39:38 am	11:57:36 am	0.803	52°
	Houston	10:43:43 am	12:02:33 pm	01:23:43 pm	0.857	66°
	San Antonio	10:39:24 am	11:56:35 am	01:17:07 pm	0.806	62°
Virginia	Richmond	12:10:38 pm	01:25:08 pm	02:37:44 pm	0.726	67°
Washington	Seattle	08:20:13 am	09:21:31 am	10:27:08 am	0.799	32°
West Virginia	Huntington	12:00:42 pm	01:16:02 pm	02:30:35 pm	0.783	66°
Wisconsin	Milwaukee	10:51:31 am	12:03:01 pm	01:15:11 pm	0.731	59°

Table 8 - Partial Eclipse Circumstances for North, Central, and South America

Country	City	Partial Eclipse Begins	Maximum Eclipse	Partial Eclipse Ends	Eclipse Magnitude	Sun Altitude
Barbados	Bridgetown	01:18:18 pm	02:34:58 pm	03:43:55 pm	0.926	53°
Bolivia	La Paz	02:21:06 pm	03:07:08 pm	03:50:14 pm	0.223	42°
Brazil	Rio de Janeiro	03:34:19 pm	04:25:38 pm	05:13:12 pm	0.459	15°
	Sao Paulo	03:35:29 pm	04:24:55 pm	05:10:54 pm	0.395	17°
Canada	Calgary, AB	09:29:16 am	10:30:06 am	11:34:13 am	0.676	37°
	Edmonton, AB	09:33:27 am	10:31:47 am	11:32:52 am	0.596	37°
	Halifax, NS	01:40:01 pm	02:34:42 pm	03:27:35 pm	0.347	56°
	Hamilton, ON	12:05:37 pm	01:14:38 pm	02:22:52 pm	0.619	61°
	Kitchener, ON	12:04:29 pm	01:13:30 pm	02:21:52 pm	0.623	61°
	Laval, QC	12:18:06 pm	01:20:31 pm	02:21:40 pm	0.477	59°
	London, ON	12:03:00 pm	01:12:54 pm	02:22:13 pm	0.645	61°
	Mississauga, ON	12:06:08 pm	01:14:37 pm	02:22:19 pm	0.607	61°
	Montreal, QC	12:18:27 pm	01:20:50 pm	02:21:56 pm	0.476	59°
	North York, ON	12:06:33 pm	01:14:42 pm	02:22:05 pm	0.600	61°
	Oshawa, ON	12:07:39 pm	01:15:22 pm	02:22:14 pm	0.589	61°
	Ottawa, ON	12:14:10 pm	01:18:06 pm	02:20:56 pm	0.508	59°
	Quebec, QC	12:23:41 pm	01:22:22 pm	02:19:45 pm	0.414	57°
	St. Catharines, ON	12:06:44 pm	01:15:35 pm	02:23:33 pm	0.613	61°
	Saskatoon, SK	10:36:29 am	11:37:24 am	12:40:51 pm	0.615	42°
	Scarborough, ON	12:06:49 pm	01:14:57 pm	02:22:16 pm	0.599	61°
	Toronto, ON	12:06:35 pm	01:14:53 pm	02:22:22 pm	0.602	61°
	Vancouver, BC	08:21:58 am	09:21:54 am	10:25:50 am	0.746	31°
	Victoria, BC	08:20:46 am	09:21:12 am	10:25:49 am	0.773	31°
	Windsor, ON	11:59:39 am	01:10:57 pm	02:21:56 pm	0.688	62°
	Winnipeg, MB	10:43:27 am	11:47:31 am	12:53:25 pm	0.624	49°
Colombia	Bogota	12:13:21 pm	01:30:03 pm	02:39:50 pm	0.622	66°
Costa Rica	San Jose	10:44:36 am	12:00:59 pm	01:12:54 pm	0.516	83°
Cuba	Havana	12:15:23 pm	01:39:11 pm	02:59:36 pm	0.892	81°
Ecuador	Quito	12:24:13 pm	01:31:16 pm	02:33:07 pm	0.399	67°
El Salvador	San Salvador	10:25:54 am	11:41:52 am	12:55:30 pm	0.501	85°
Guatemala	Guatemala	10:21:13 am	11:36:54 am	12:50:54 pm	0.500	83°
Haiti	Port-au-Prince	01:43:18 pm	03:06:23 pm	04:22:53 pm	0.994	72°
Honduras	Tegucigalpa	10:27:37 am	11:45:45 am	01:00:49 pm	0.554	88°
Jamaica	Kingston	11:35:43 am	12:59:34 pm	02:17:36 pm	0.882	78°
Mexico	Mexico City	10:55:25 am	12:07:11 pm	01:21:22 pm	0.487	67°
	Guadalajara	10:46:30 am	11:55:16 am	01:07:58 pm	0.473	60°
	Monterrey	10:41:25 am	11:56:37 am	01:15:29 pm	0.666	62°
Nicaragua	Managua	10:34:45 am	11:51:43 am	01:05:05 pm	0.523	87°
Panama	Panama	11:53:39 am	01:12:09 pm	02:24:47 pm	0.598	77°
Paraguay	Asuncion	03:43:22 pm	04:20:35 pm	04:55:56 pm	0.176	26°
Peru	Lima	01:10:53 pm	01:51:26 pm	02:29:55 pm	0.142	54°
Puerto Rico	San Juan	12:56:05 pm	02:16:18 pm	03:29:23 pm	0.946	64°
Venezuela	Caracas	01:09:46 pm	02:30:03 pm	03:42:27 pm	0.929	61°

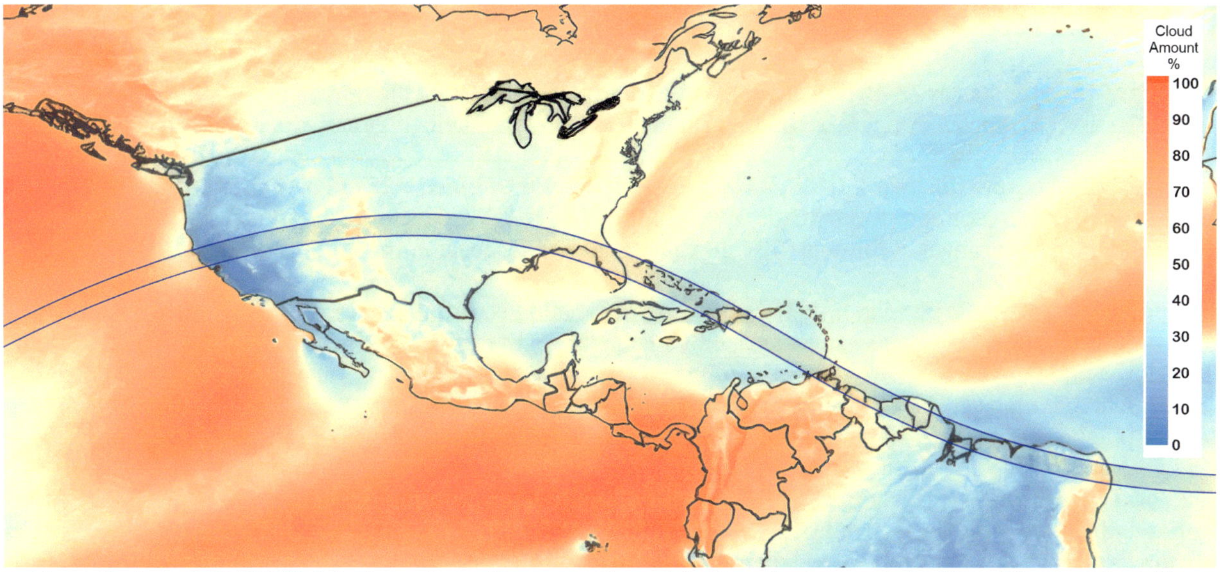

Figure 9 – August 12, 2045 total solar eclipse with average August daytime cloud cover derived from 20 years of satellite data, 2000-2019. Data: CM-SAF. (based on map by Jay Anderson)

5 – Average Cloud Cover for the 2045 Total Solar Eclipse

The average cloud cover along the 2045 total eclipse track is lowest in the western USA. It increases in the eastern USA peaking in Florida. Some locations in the West Indies and South America also offer fair prospects.

Table 9 – Average Cloud Cover and High Temperature in August

State/Country	City	Duration Totality	Ave. High Temp.[1]	Ave. Cloud Cover[2]
California	Redding	04m25s	36°C (96°F)	9%
California	Susanville	04m33s	30°C (84°F)	10%
Nevada	Reno	02m45s	31°C (88°F)	9%
Nevada	Winnemucca	03m28s	32°C (89°F)	12%
Utah	Salt Lake City	01m32s	32°C (89°F)	20%
Utah	Moab	03m25s	32°C (90°F)	29%
Colorado	Grand Junction	05m04s	32°C (89°F)	30%
Colorado	Colorado Springs	05m04s	27°C (80°F)	40%
Oklahoma	Enid	05m34s	33°C (92°F)	45%
Oklahoma	Tulsa	05m33s	33°C (92°F)	44%
Arkansas	Fort Smith	05m44s	33°C (92°F)	39%
Arkansas	Little Rock	05m36s	33°C (91°F)	58%
Mississippi	Jackson	03m10s	32°C (90°F)	61%
Mississippi	Meridian	05m51s	32°C (89°F)	59%
Alabama	Dothan	05m40s	31°C (88°F)	71%
Florida	Panama City	05m10s	31°C (87°F)	69%
Florida	Tallahassee	05m47s	32°C (89°F)	65%
Florida	Gainesville	04m36s	31°C (88°F)	68%
Florida	Orlando	05m42s	32°C (89°F)	72%
Florida	Miami	02m42s	31°C (88°F)	74%
Bahamas	Freeport	05m47s	30°C (86°F)	60%
Bahamas	Nassau	06m01s	31°C (88°F)	63%
Dominican Republic	Santo Domingo	05m40s	31°C (88°F)	80%
Trinidad & Tabago	San Fernando	05m11s	31°C (87°F)	80%
Guyana	Georgetown	04m56s	30°C (86°F)	50%
Suriname	Paramaribo	02m33s	31°C (88°F)	40%
Brazil	Belem	01m30s	32°C (89°F)	49%
Brazil	Sao Luis	03m57s	31°C (88°F)	41%
Brazil	Caico	03m39s	35°C (95°F)	35%
Brazil	Joao Pessoa	03m26s	28°C (82°F)	70%

[1]*Average High Temperature in August [source: www.weatherspark.com]*

[2]*Average Cloud Cover in August (since 2000) [source: www.timeanddate.com]*

6 – Detailed Maps of the Path of Totality

A series of 19 maps (pages 22-40) cover the 2045 path of totality through the USA[6]. The *Table of Contents* (page iii) can be used to quickly navigate to the map of interest since it lists the region featured on each map.

The map scale is approximately 1:1,400,000, which corresponds to 1 inch ≈ 22 miles (1 cm ≈ 14 km).[7] This large scale shows both major and minor roads, towns and cities, rivers, lakes, parks, national forests, wilderness areas and mountain ranges. A 50-mile (50-km) reference scale appears at the bottom of each map.

The path of totality on each map is depicted as a lightly shaded region with the northern and southern limits clearly labeled. The total eclipse can be seen only inside this path (a partial eclipse is visible outside the path). The closer one gets to the central line, the longer the total eclipse lasts. Gray lines inside the path mark the duration of the total eclipse in 30-second steps. This makes it easy to estimate the duration of totality from any location in the eclipse path.

The local time of mid-eclipse is marked by a series of white lines crossing the eclipse path every 5 minutes. Abbreviations for local times are: PDT, MDT, CDT, EDT, AST, VST, GYT, SRT, and BRT (see Table 1). Eclipse circumstances on the central line are labeled with the local time of mid-eclipse, duration of totality (minutes and seconds) and altitude of the Sun.

All maps were produced using Google Maps as the underlying map with overlying eclipse graphics generated using JavaScript code. A web page is available to the user for examining any part of the 2045 eclipse path at a range of zoom magnifications. An added benefit of the web page is that it automatically calculates the local circumstances for any point chosen by the user, with just the click of a mouse. For more information and to access the interactive 2045 eclipse path plotted on Google Maps, visit:

https://www.eclipsewise.com/solar/SEgmapx/2001-2100/SE2045Aug12Tgmapx.html#map

7 – EclipseWise.com Web Site

EclipseWise.com has individual web pages, maps and diagrams for every solar and lunar eclipse covering the 5000-year period from 2000 BCE to 3000 CE. This includes 11,898 solar eclipses and 12,064 lunar eclipses. Much of the design, layout and graphics were inspired by the recent publications *Thousand Year Canon of Solar Eclipses 1501 to 2500* and the *Thousand Year Canon of Lunar Eclipses 1501 to 2500.* For more information about these two books, visit: ***https://www.astropixels.com/pubs***

The graphical user interface used by *EclipseWise.com* offers an intuitive way of accessing eclipse predictions. For example, the home page presents a concise preview of all upcoming solar and lunar eclipses over several years. Each small eclipse diagram gives a quick preview of an eclipse and links to a dedicated page for that eclipse.

The main or top pages of EclipseWise.com are:

Home Page (both solar and lunar eclipses): ***https://www. eclipsewise.com/eclipse.html***
Solar Eclipses Page: ***https://www. eclipsewise.com/solar/solar.html***
Lunar Eclipses Page: ***https://www. eclipsewise.com/lunar/lunar.html***

---

6 The **International Edition** includes an additional 18 full-page maps of the path through the West Indies and South America.

7 Because of the Mercator map projection, the actual scale on a given map can vary by up to 10% of this value.

8 – EclipseWise.com and the 2045 Eclipse

EclipseWise.com has a series of pages and resources devoted to the 2045 eclipse. The main page is located at:

https://www.eclipsewise.com/solar/SEprime/2001-2100/SE2045Aug12Tprime.html

It provides links to detailed eclipse path maps, tables of eclipse path coordinates, Besselian elements, and more. The link to an interactive Google Map with the eclipse path plotted on it allows the user to zoom into a part of the path. Click on any point on the map to display the eclipse circumstances and duration of totality at that location.

Other features include information on eye safety, eclipse photography, the sky during totality, and additional data tables about the eclipse path. This web site will continue to add features as the eclipse approaches.

The *2045 Total Solar Eclipse Circumstances Calculator* is an interactive web page that can quickly calculate the local circumstances for the eclipse from any geographic location. The *Calculator* is located at:

https://www.eclipsewise.com/solar/SEcirc/2001-2100/SE2045Aug12Tcirc.html

9 – Eclipse Predictions

The algorithms and software for the eclipse predictions were developed primarily from the *Explanatory Supplement to the Astronomical Ephemeris* (Her Majesty's Nautical Almanac Office, 1974) with additional algorithms from *Elements of Solar Eclipses: 1951–2200* (Meeus, 1989). The solar and lunar ephemerides were generated from the JPL DE430. All eclipse calculations were made using a value for the Moon's radius of k=0.2722810 for the path of totality. Times for local circumstances in Tables 3 through 6 include the effects of the Moon's limb profile. A value for ΔT of 72.2 seconds was used to convert the predictions from Terrestrial Time (TT) to Coordinated Universal Time (UTC).

Photo 3 – Viewing the Total Solar Eclipse of August 1, 2008 from Jinta, China.
© 2008 F. Espenak, www.MrEclipse.com

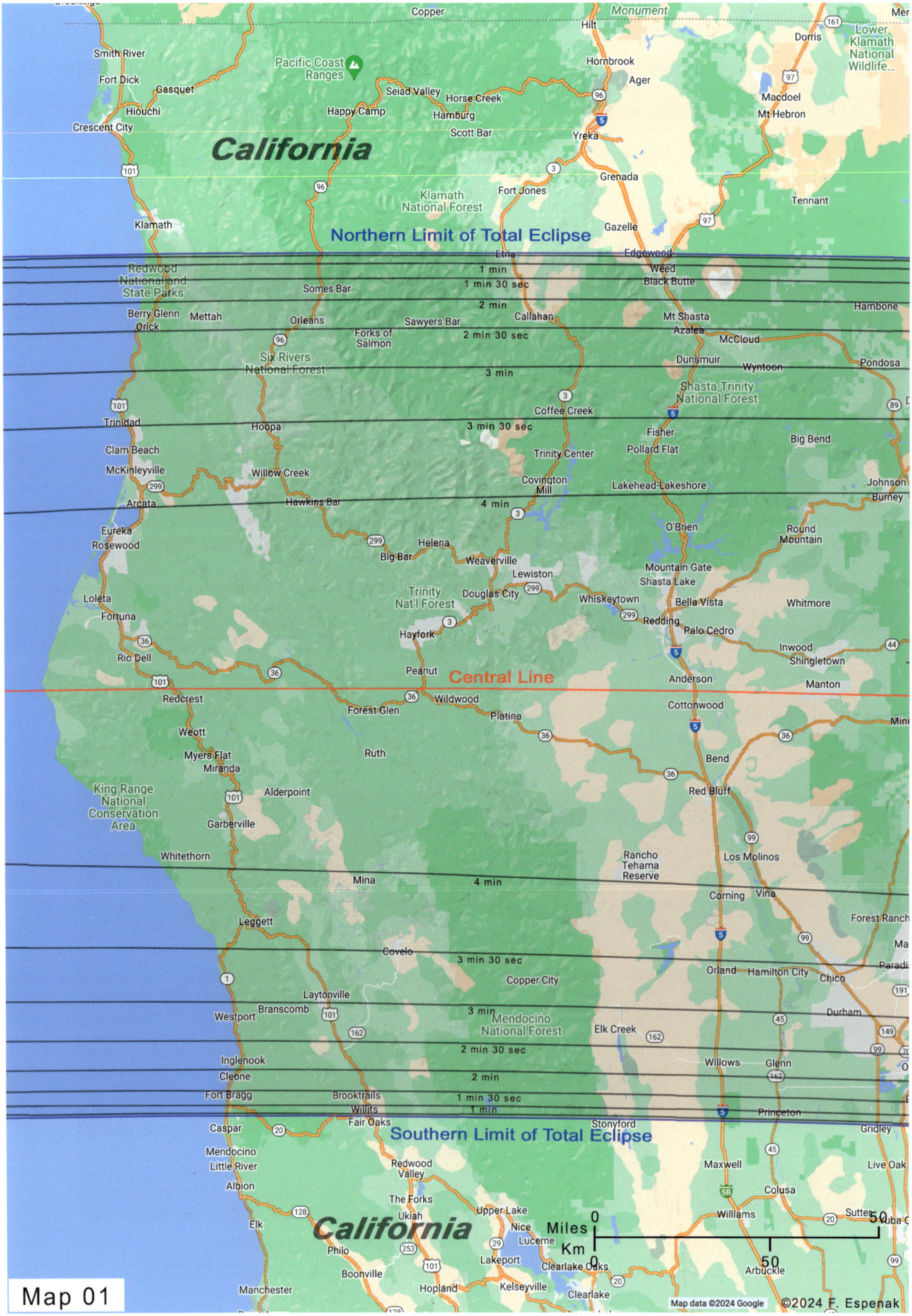
California
Northern Limit of Total Eclipse
1 min
1 min 30 sec
2 min
2 min 30 sec
3 min
3 min 30 sec
4 min
Central Line
4 min
3 min 30 sec
3 min
2 min 30 sec
2 min
1 min 30 sec
1 min
Southern Limit of Total Eclipse
California
Smith River
Fort Dick
Gasquet
Hiouchi
Crescent City
Copper
Pacific Coast Ranges
Seiad Valley
Happy Camp
Horse Creek
Hamburg
Scott Bar
Monument
Hilt
Hornbrook
Ager
Yreka
Grenada
Fort Jones
Klamath National Forest
Gazelle
Dorris
Lower Klamath National Wildlife...
Macdoel
Mt Hebron
Tennant
Klamath
Etna
Edgewood
Weed
Black Butte
Redwood National and State Parks
Somes Bar
Berry Glenn
Orick
Mettah
Orleans
Forks of Salmon
Sawyers Bar
Callahan
Mt Shasta
Azalea
McCloud
Hambone
Six Rivers National Forest
Dunsmuir
Wyntoon
Pondosa
Shasta-Trinity National Forest
Coffee Creek
Trinidad
Hoopa
Fisher
Pollard Flat
Big Bend
Clam Beach
McKinleyville
Willow Creek
Trinity Center
Covington Mill
Lakehead-Lakeshore
Johnson
Burney
Arcata
Hawkins Bar
Eureka
Rosewood
O'Brien
Round Mountain
Helena
Big Bar
Weaverville
Lewiston
Mountain Gate
Shasta Lake
Trinity Nat'l Forest
Douglas City
Whiskeytown
Bella Vista
Whitmore
Loleta
Fortuna
Redding
Palo Cedro
Hayfork
Inwood
Shingletown
Rio Dell
Peanut
Anderson
Manton
Redcrest
Wildwood
Forest Glen
Platina
Cottonwood
Weott
Ruth
Bend
Myers Flat
Miranda
King Range National Conservation Area
Alderpoint
Red Bluff
Garberville
Whitethorn
Rancho Tehama Reserve
Los Molinos
Mina
Corning
Vina
Leggett
Forest Ranch
Covelo
Copper City
Orland
Hamilton City
Chico
Laytonville
Westport
Branscomb
Mendocino National Forest
Elk Creek
Durham
Inglenook
Cleone
Willows
Glenn
Fort Bragg
Brooktrails
Willits
Princeton
Caspar
Fair Oaks
Stonyford
Gridley
Mendocino
Little River
Albion
Redwood Valley
Maxwell
Live Oak
The Forks
Colusa
Ukiah
Upper Lake
Williams
Sutter
Elk
Nice
Lucerne
Philo
Lakeport
Clearlake Oaks
Arbuckle
Boonville
Manchester
Hopland
Kelseyville
Clearlake
Miles 0 50
Km 0 50
Map 01
Map data ©2024 Google
©2024 F. Espenak

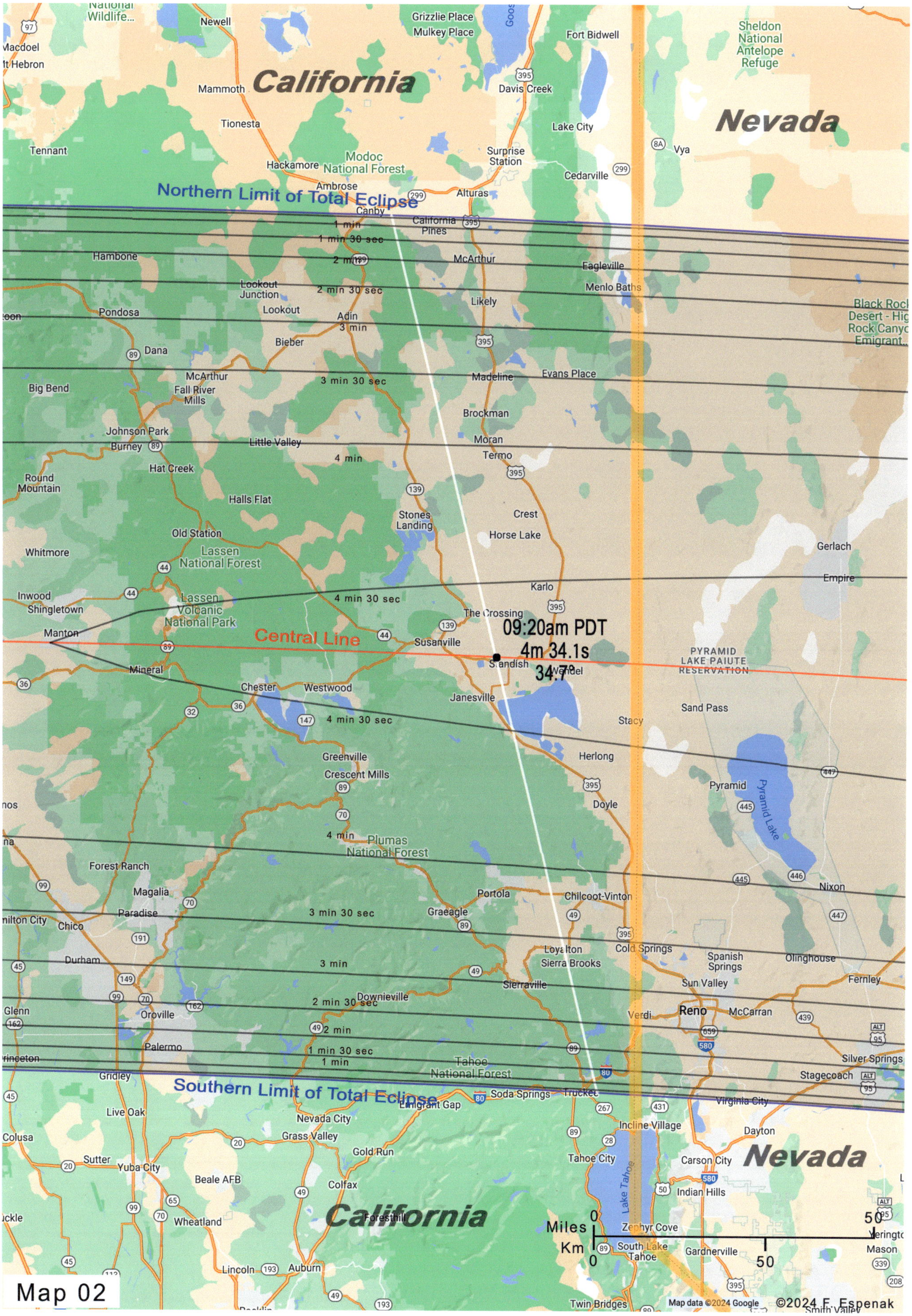
California
Nevada
Northern Limit of Total Eclipse
1 min
1 min 30 sec
2 min
2 min 30 sec
3 min
3 min 30 sec
4 min
4 min 30 sec
Central Line
09:20am PDT
4m 34.1s
34.7°
4 min 30 sec
4 min
3 min 30 sec
3 min
2 min 30 sec
2 min
1 min 30 sec
1 min
Southern Limit of Total Eclipse
Nevada
California
Miles
Km
0
50
Map data ©2024 Google
©2024 F. Espenak
Map 02

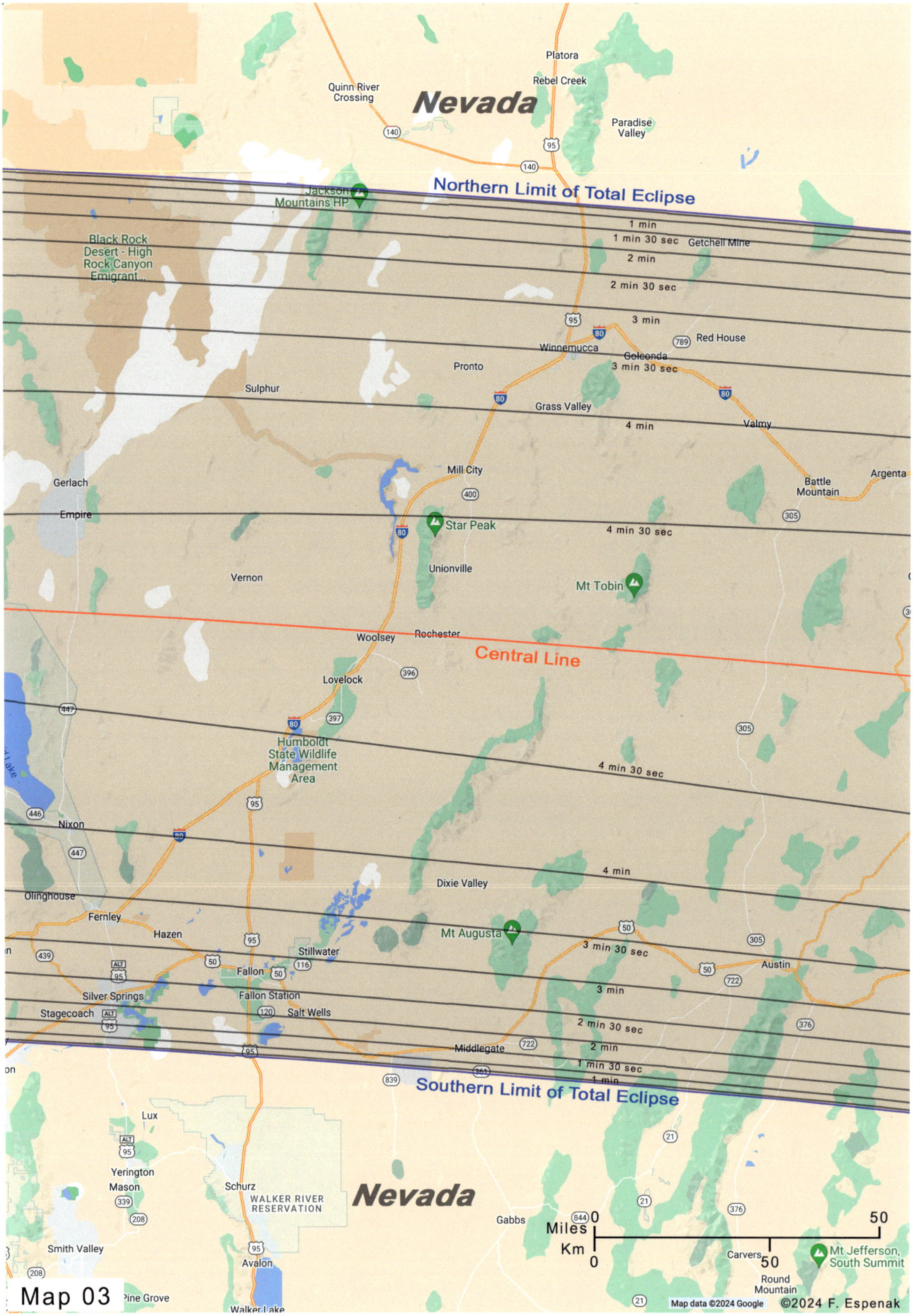
Nevada
Northern Limit of Total Eclipse
Central Line
Southern Limit of Total Eclipse
1 min
1 min 30 sec
2 min
2 min 30 sec
3 min
3 min 30 sec
4 min
4 min 30 sec
Quinn River Crossing
Platora
Rebel Creek
Paradise Valley
Jackson Mountains HP
Getchell Mine
Black Rock Desert - High Rock Canyon Emigrant...
Winnemucca
Red House
Golconda
Pronto
Sulphur
Grass Valley
Valmy
Mill City
Battle Mountain
Argenta
Gerlach
Empire
Star Peak
Unionville
Vernon
Mt Tobin
Woolsey
Rochester
Lovelock
Humboldt State Wildlife Management Area
Nixon
Olinghouse
Fernley
Hazen
Dixie Valley
Mt Augusta
Stillwater
Fallon
Fallon Station
Salt Wells
Silver Springs
Stagecoach
Austin
Middlegate
Lux
Yerington
Mason
Schurz
WALKER RIVER RESERVATION
Gabbs
Smith Valley
Avalon
Walker Lake
Carvers
Round Mountain
Mt Jefferson, South Summit
Miles 0 50
Km 0 50
Map 03
Map data ©2024 Google
©2024 F. Espenak

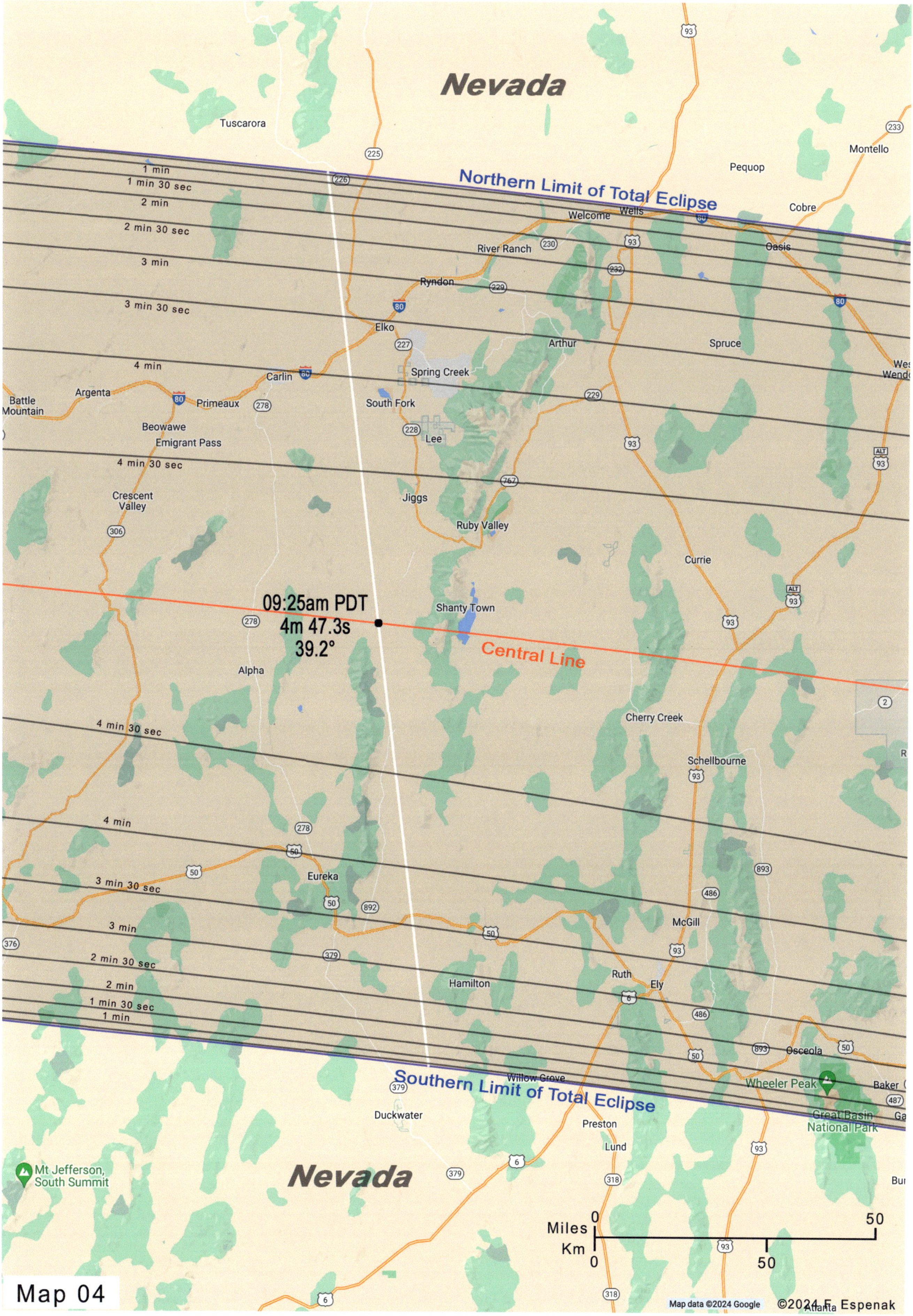
Nevada
Tuscarora
Northern Limit of Total Eclipse
1 min
1 min 30 sec
2 min
2 min 30 sec
3 min
3 min 30 sec
4 min
4 min 30 sec
Pequop
Montello
Cobre
Welcome
Wells
Oasis
River Ranch
Ryndon
Elko
Arthur
Spruce
Spring Creek
Carlin
Argenta
Battle Mountain
Primeaux
South Fork
Beowawe
Emigrant Pass
Lee
Crescent Valley
Jiggs
Ruby Valley
Currie
09:25am PDT
4m 47.3s
39.2°
Shanty Town
Central Line
Alpha
Cherry Creek
Schellbourne
Eureka
McGill
Hamilton
Ruth
Ely
Osceola
Willow Grove
Southern Limit of Total Eclipse
Wheeler Peak
Baker
Great Basin National Park
Duckwater
Preston
Lund
Mt Jefferson, South Summit
Miles 0 50
Km 0 50
Map 04
Map data ©2024 Google
©2024 F. Espenak

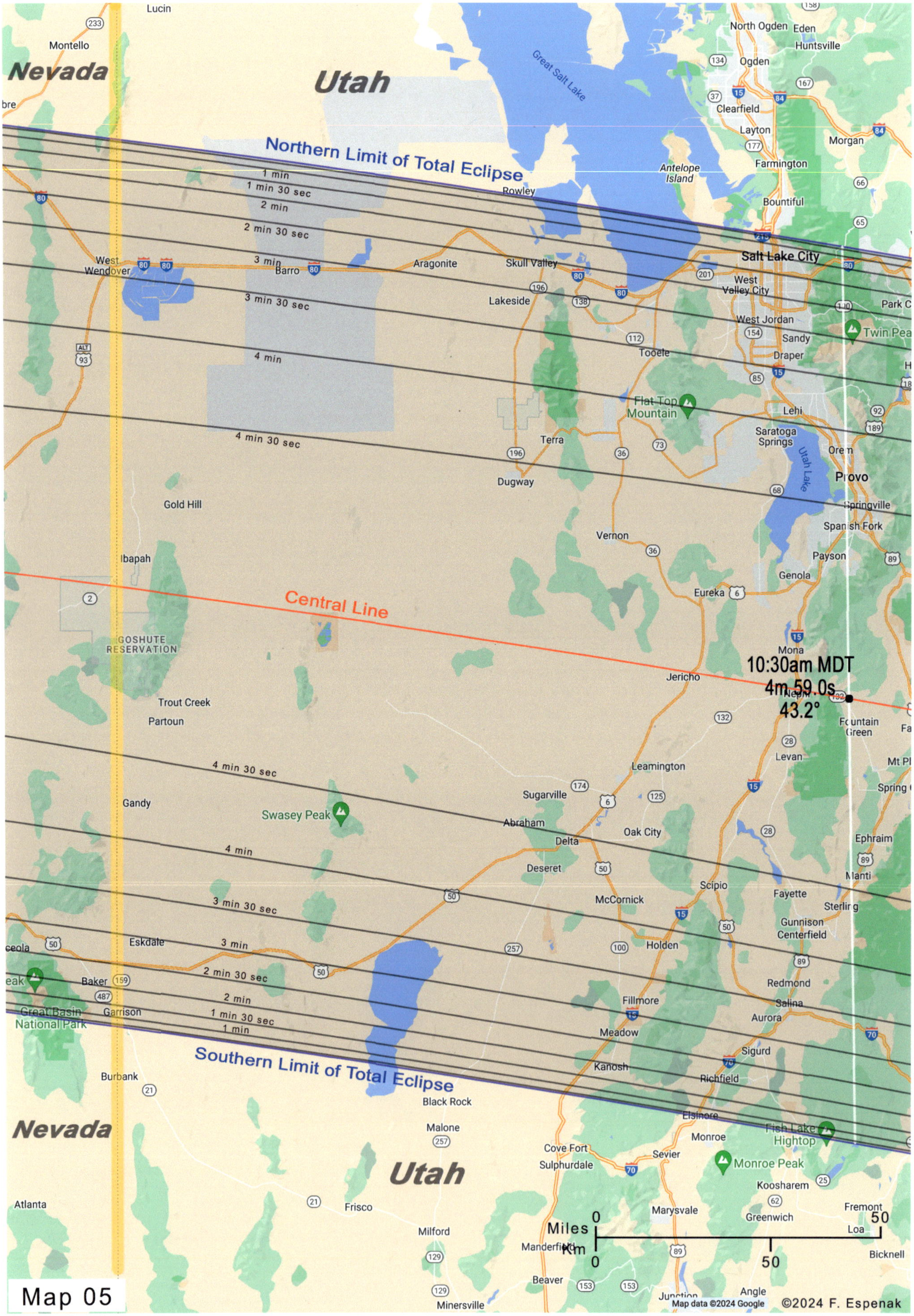
Nevada
Utah
Great Salt Lake
Northern Limit of Total Eclipse
1 min
1 min 30 sec
2 min
2 min 30 sec
3 min
3 min 30 sec
4 min
4 min 30 sec
Central Line
10:30am MDT
4m 59.0s
43.2°
4 min 30 sec
4 min
3 min 30 sec
3 min
2 min 30 sec
2 min
1 min 30 sec
1 min
Southern Limit of Total Eclipse
Nevada
Utah
Lucin
Montello
North Ogden
Eden
Huntsville
Ogden
Clearfield
Layton
Morgan
Farmington
Antelope Island
Bountiful
Rowley
Salt Lake City
West Wendover
Barro
Aragonite
Skull Valley
Lakeside
West Valley City
West Jordan
Sandy
Draper
Tooele
Flat Top Mountain
Lehi
Saratoga Springs
Utah Lake
Provo
Springville
Spanish Fork
Payson
Genola
Terra
Dugway
Gold Hill
Vernon
Ibapah
Eureka
GOSHUTE RESERVATION
Mona
Jericho
Nephi
Fountain Green
Trout Creek
Partoun
Levan
Leamington
Gandy
Swasey Peak
Sugarville
Abraham
Delta
Oak City
Deseret
McCornick
Scipio
Fayette
Ephraim
Manti
Sterling
Gunnison
Centerfield
Eskdale
Holden
Baker
Great Basin National Park
Garrison
Fillmore
Redmond
Salina
Aurora
Meadow
Kanosh
Sigurd
Richfield
Burbank
Black Rock
Elsinore
Malone
Monroe
Fish Lake Hightop
Cove Fort
Sulphurdale
Sevier
Monroe Peak
Koosharem
Atlanta
Frisco
Marysvale
Greenwich
Fremont
Loa
Milford
Manderfield
Bicknell
Beaver
Junction
Angle
Minersville
Miles 0 50
Km 0 50
Map data ©2024 Google
©2024 F. Espenak
Map 05

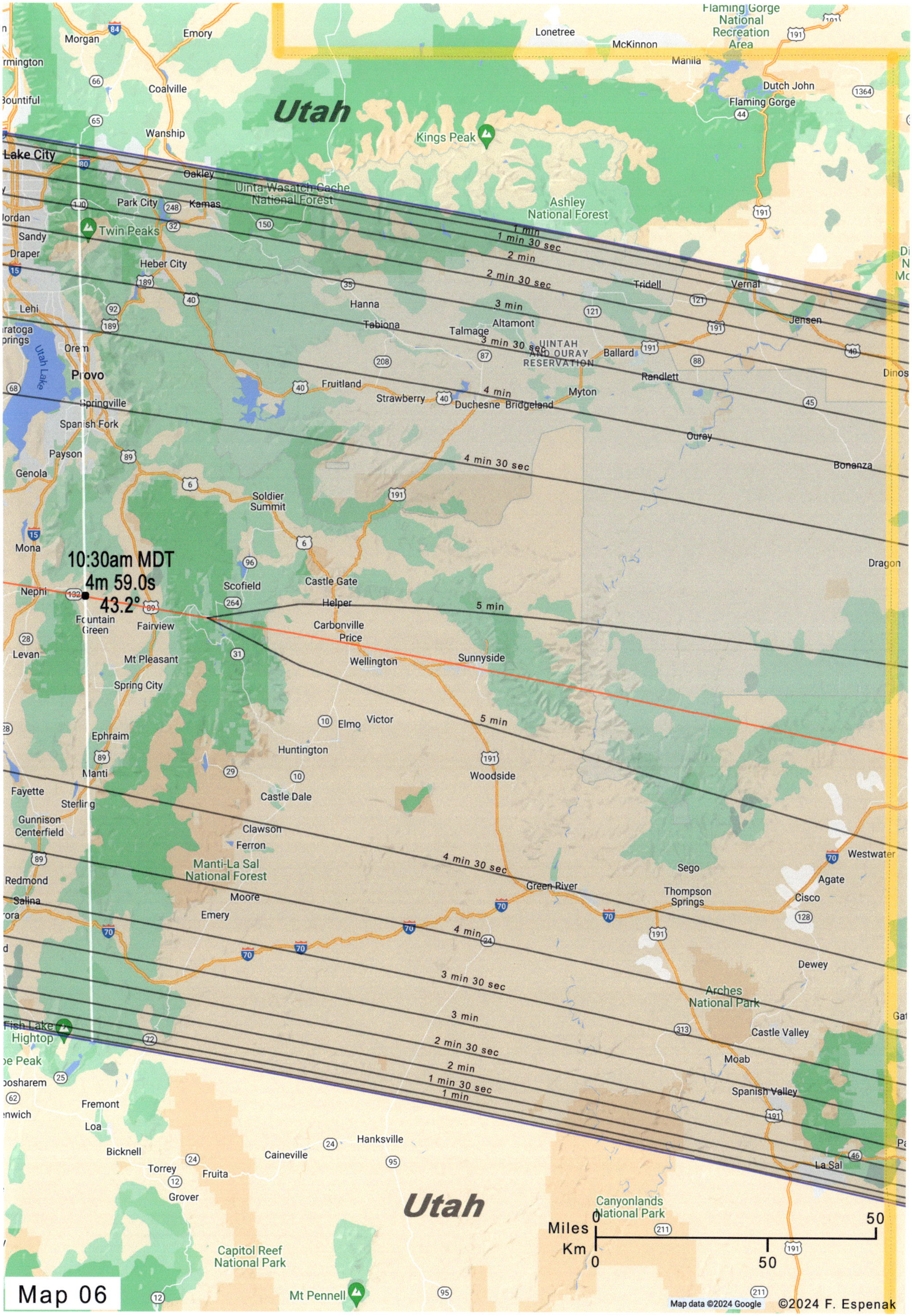

Utah
Morgan
Emory
Lonetree
McKinnon
Manila
Flaming Gorge National Recreation Area
Dutch John
Flaming Gorge
Coalville
Bountiful
Wanship
Kings Peak
Lake City
Oakley
Uinta-Wasatch-Cache National Forest
Ashley National Forest
Park City
Kamas
Jordan
Sandy
Twin Peaks
Draper
Heber City
1 min
1 min 30 sec
2 min
2 min 30 sec
3 min
3 min 30 sec
4 min
4 min 30 sec
5 min
Tridell
Vernal
Jensen
Lehi
Hanna
Tabiona
Altamont
Talmage
UINTAH AND OURAY RESERVATION
Ballard
Randlett
Utah Lake
Orem
Provo
Fruitland
Strawberry
Duchesne
Bridgeland
Myton
Springville
Spanish Fork
Ouray
Payson
Bonanza
Genola
Soldier Summit
Mona
10:30am MDT
4m 59.0s
43.2°
Dragon
Nephi
Scofield
Castle Gate
Helper
Fountain Green
Fairview
Carbonville
Price
Levan
Mt Pleasant
Wellington
Sunnyside
Spring City
Elmo
Victor
Ephraim
Huntington
Woodside
Manti
Fayette
Castle Dale
Sterling
Gunnison
Centerfield
Clawson
Ferron
Manti-La Sal National Forest
Westwater
Sego
Agate
Redmond
Green River
Thompson Springs
Salina
Moore
Emery
Cisco
Dewey
Arches National Park
Castle Valley
Fish Lake
Hightop
Moab
Spanish Valley
Fremont
Loa
Bicknell
Torrey
Fruita
Grover
Caineville
Hanksville
La Sal
Canyonlands National Park
Miles
Km
0
50
Capitol Reef National Park
Mt Pennell
Map 06
Map data ©2024 Google
©2024 F. Espenak

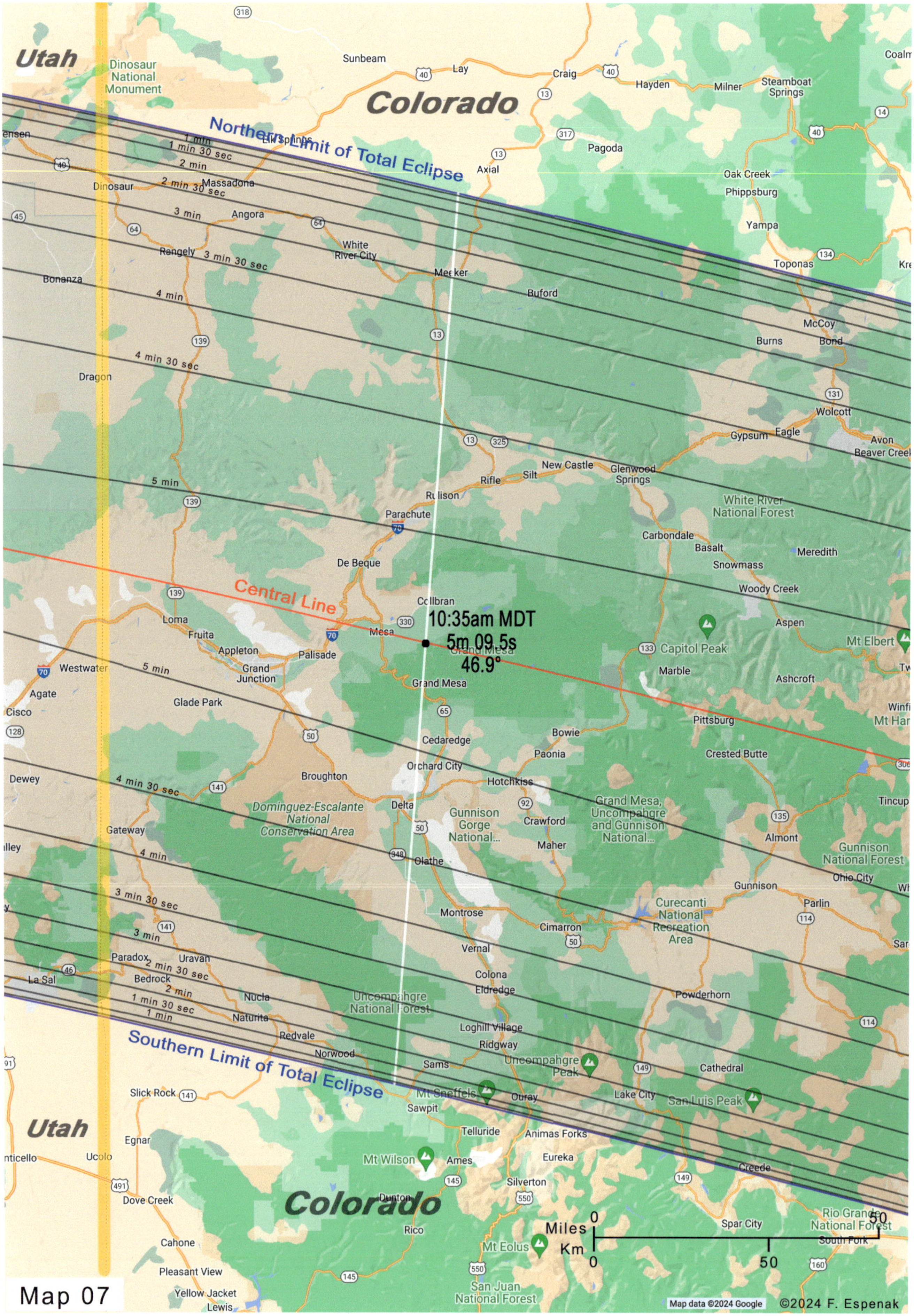
Utah
Colorado
Northern Limit of Total Eclipse
1 min
1 min 30 sec
2 min
2 min 30 sec
3 min
3 min 30 sec
4 min
4 min 30 sec
5 min
Central Line
10:35am MDT
5m 09.5s
46.9°
5 min
4 min 30 sec
4 min
3 min 30 sec
3 min
2 min 30 sec
2 min
1 min 30 sec
1 min
Southern Limit of Total Eclipse
Utah
Colorado
Dinosaur National Monument
Sunbeam
Lay
Craig
Hayden
Milner
Steamboat Springs
Pagoda
Axial
Oak Creek
Phippsburg
Yampa
Toponas
Dinosaur
Massadona
Angora
Rangely
White River City
Meeker
Buford
Bonanza
McCoy
Burns
Bond
Dragon
Wolcott
Gypsum
Eagle
Avon
Beaver Creek
New Castle
Glenwood Springs
Silt
Rifle
Rulison
Parachute
White River National Forest
Carbondale
Basalt
Meredith
Snowmass
De Beque
Woody Creek
Collbran
Aspen
Loma
Fruita
Mesa
Capitol Peak
Mt Elbert
Appleton
Palisade
Grand Junction
Westwater
Marble
Ashcroft
Agate
Grand Mesa
Glade Park
Cisco
Pittsburg
Cedaredge
Bowie
Paonia
Crested Butte
Orchard City
Dewey
Broughton
Hotchkiss
Dominguez-Escalante National Conservation Area
Delta
Gunnison Gorge National...
Crawford
Grand Mesa, Uncompahgre and Gunnison National...
Almont
Gateway
Maher
Olathe
Gunnison National Forest
Ohio City
Gunnison
Montrose
Curecanti National Recreation Area
Parlin
Cimarron
Vernal
Paradox
Uravan
Bedrock
Colona
La Sal
Eldredge
Nucla
Naturita
Uncompahgre National Forest
Powderhorn
Loghill Village
Redvale
Ridgway
Norwood
Sams
Uncompahgre Peak
Cathedral
Mt Sneffels
Ouray
Lake City
San Luis Peak
Slick Rock
Sawpit
Telluride
Animas Forks
Egnar
Eureka
Ucolo
Mt Wilson
Ames
Creede
Silverton
Dove Creek
Dunton
Rico
Rio Grande National Forest
Spar City
South Fork
Mt Eolus
Cahone
Pleasant View
Yellow Jacket
Lewis
San Juan National Forest
Miles 0 50
Km 0 50
Map 07
Map data ©2024 Google
©2024 F. Espenak

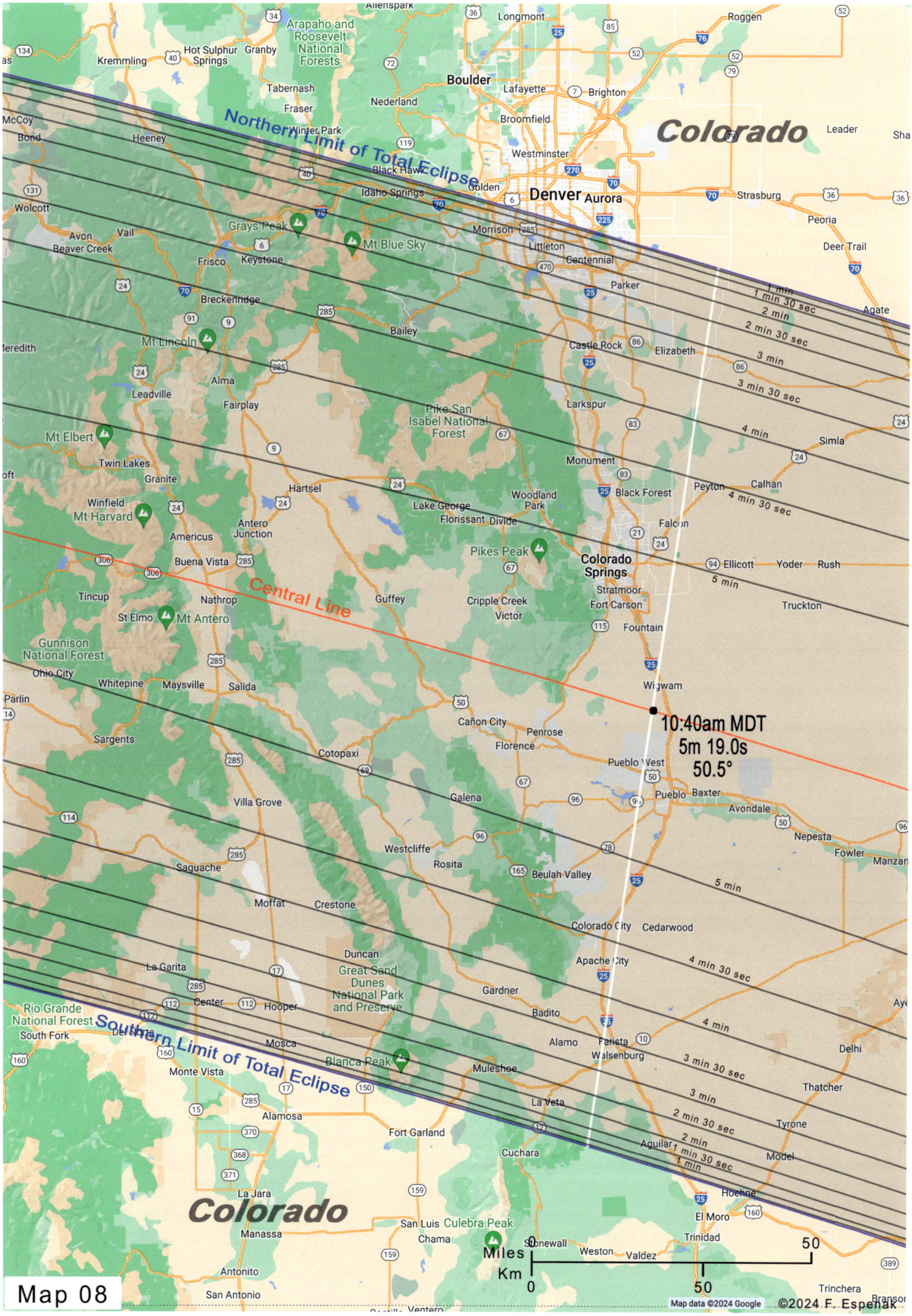
Map 08
Northern Limit of Total Eclipse
Central Line
Southern Limit of Total Eclipse
10:40am MDT
5m 19.0s
50.5°
1 min
1 min 30 sec
2 min
2 min 30 sec
3 min
3 min 30 sec
4 min
4 min 30 sec
5 min
Colorado
Denver
Aurora
Boulder
Colorado Springs
Pueblo
Pueblo West
Walsenburg
Trinidad
Longmont
Lafayette
Brighton
Broomfield
Westminster
Golden
Idaho Springs
Black Hawk
Morrison
Littleton
Centennial
Parker
Castle Rock
Elizabeth
Larkspur
Monument
Black Forest
Falcon
Peyton
Calhan
Simla
Ellicott
Yoder
Rush
Truckton
Stratmoor
Fort Carson
Fountain
Wigwam
Cañon City
Penrose
Florence
Baxter
Avondale
Nepesta
Fowler
Beulah Valley
Colorado City
Cedarwood
Apache City
Gardner
Badito
Alamo
Farista
La Veta
Cuchara
Aguilar
Hoehne
El Moro
Stonewall
Weston
Valdez
Kremmling
Hot Sulphur Springs
Granby
Tabernash
Fraser
Winter Park
Nederland
Heeney
Wolcott
Avon
Vail
Beaver Creek
Frisco
Keystone
Breckenridge
Grays Peak
Mt Blue Sky
Mt Lincoln
Alma
Fairplay
Leadville
Bailey
Mt Elbert
Twin Lakes
Granite
Winfield
Mt Harvard
Americus
Antero Junction
Hartsel
Buena Vista
Nathrop
Tincup
St Elmo
Mt Antero
Gunnison National Forest
Ohio City
Whitepine
Maysville
Salida
Parlin
Sargents
Cotopaxi
Villa Grove
Saguache
Moffat
Crestone
Westcliffe
Rosita
Galena
Guffey
Cripple Creek
Victor
Pikes Peak
Woodland Park
Divide
Florissant
Lake George
Pike-San Isabel National Forest
Arapaho and Roosevelt National Forests
La Garita
Center
Hooper
Duncan
Great Sand Dunes National Park and Preserve
Mosca
Blanca Peak
Mulleshoe
Fort Garland
Rio Grande National Forest
South Fork
Del Norte
Monte Vista
Alamosa
La Jara
Manassa
Antonito
San Antonio
San Luis
Chama
Culebra Peak
Strasburg
Peoria
Deer Trail
Agate
Leader
Roggen
Delhi
Thatcher
Tyrone
Model
Trinchera
Miles 0 50
Km 0 50
Map data ©2024 Google
©2024 F. Espenak

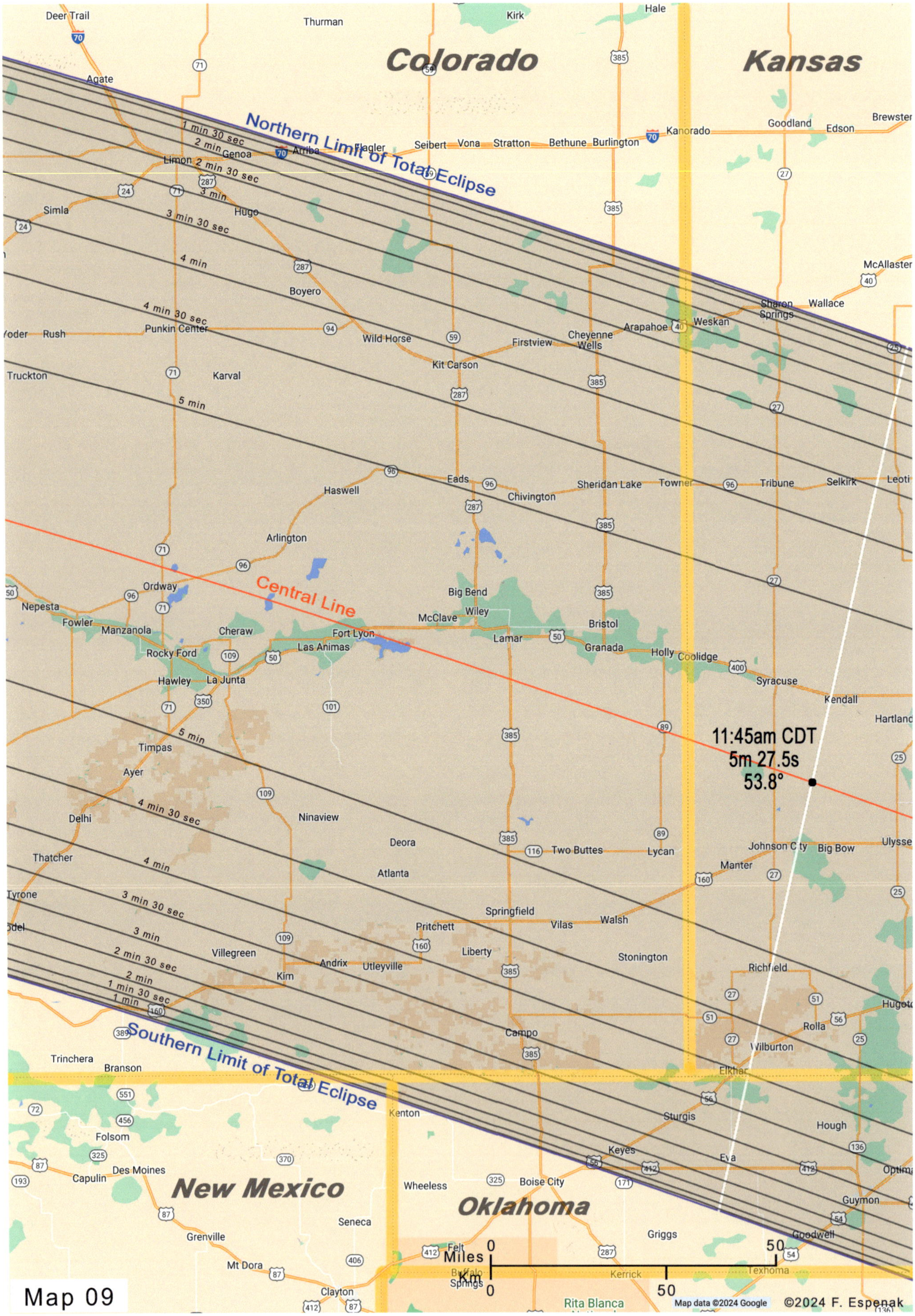
Colorado
Kansas
New Mexico
Oklahoma
Northern Limit of Total Eclipse
Central Line
Southern Limit of Total Eclipse
1 min 30 sec
2 min
2 min 30 sec
3 min
3 min 30 sec
4 min
4 min 30 sec
5 min
5 min
4 min 30 sec
4 min
3 min 30 sec
3 min
2 min 30 sec
2 min
1 min 30 sec
1 min
11:45am CDT
5m 27.5s
53.8°
Deer Trail
Thurman
Kirk
Hale
Agate
Genoa
Arriba
Flagler
Seibert
Vona
Stratton
Bethune
Burlington
Kanorado
Goodland
Edson
Brewster
Limon
Hugo
Simla
Boyero
McAllaster
Sharon Springs
Wallace
Weskan
Arapahoe
Cheyenne Wells
Firstview
Kit Carson
Wild Horse
Punkin Center
Rush
Yoder
Truckton
Karval
Eads
Haswell
Chivington
Sheridan Lake
Towner
Tribune
Selkirk
Leoti
Arlington
Ordway
Nepesta
Fowler
Manzanola
Cheraw
Rocky Ford
Fort Lyon
Las Animas
Big Bend
Wiley
McClave
Lamar
Bristol
Granada
Holly
Coolidge
Syracuse
Kendall
Hartland
Hawley
La Junta
Timpas
Ayer
Delhi
Thatcher
Tyrone
Ninaview
Deora
Atlanta
Two Buttes
Lycan
Johnson City
Big Bow
Ulysses
Manter
Springfield
Vilas
Walsh
Pritchett
Liberty
Stonington
Villegreen
Andrix
Utleyville
Kim
Richfield
Campo
Rolla
Wilburton
Elkhart
Trinchera
Branson
Kenton
Sturgis
Hough
Keyes
Eva
Folsom
Des Moines
Capulin
Wheeless
Boise City
Optima
Guymon
Goodwell
Seneca
Grenville
Mt Dora
Clayton
Felt
Griggs
Texhoma
Kerrick
Rita Blanca
Miles
Km
0
50
0
50
Map 09
Map data ©2024 Google
©2024 F. Espenak

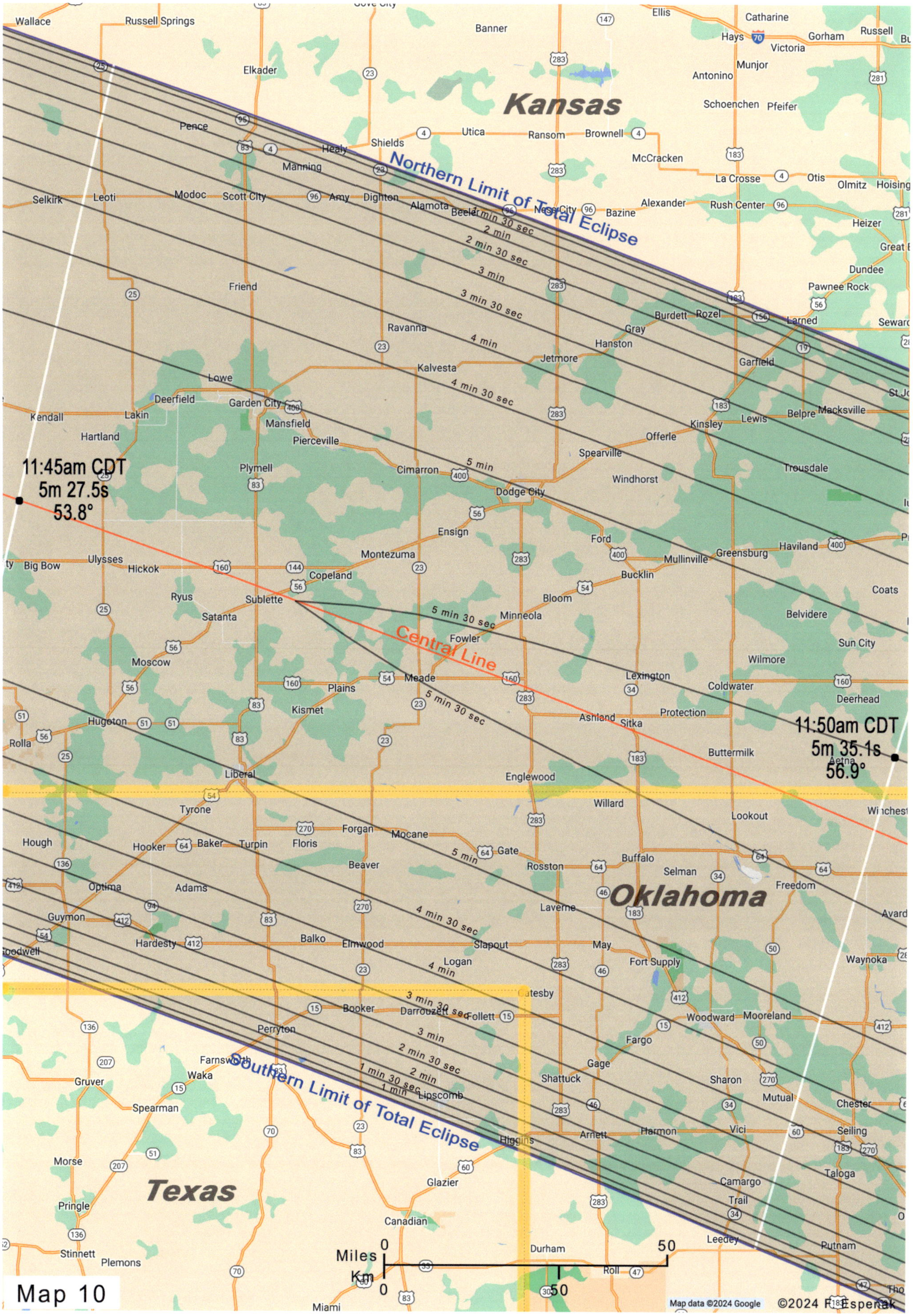
Kansas
Oklahoma
Texas
Northern Limit of Total Eclipse
Southern Limit of Total Eclipse
Central Line
1 min 30 sec
2 min
2 min 30 sec
3 min
3 min 30 sec
4 min
4 min 30 sec
5 min
5 min 30 sec
11:45am CDT
5m 27.5s
53.8°
11:50am CDT
5m 35.1s
56.9°
Miles
Km
0
50
Map 10
Map data ©2024 Google
©2024 F. Espenak
Wallace
Russell Springs
Banner
Ellis
Catharine
Hays
Gorham
Russell
Victoria
Munjor
Antonino
Elkader
Schoenchen
Pfeifer
Pence
Utica
Ransom
Brownell
Shields
Healy
Manning
McCracken
La Crosse
Otis
Olmitz
Selkirk
Leoti
Modoc
Scott City
Amy
Dighton
Alamota
Beeler
Ness City
Bazine
Alexander
Rush Center
Heizer
Friend
Dundee
Pawnee Rock
Burdett
Rozel
Larned
Gray
Ravanna
Hanston
Jetmore
Kalvesta
Garfield
Lowe
Deerfield
Garden City
Kendall
Lakin
Mansfield
Kinsley
Lewis
Belpre
Macksville
Hartland
Pierceville
Offerle
Spearville
Plymell
Cimarron
Trousdale
Windhorst
Dodge City
Ensign
Ford
Haviland
Greensburg
Mullinville
Big Bow
Ulysses
Hickok
Montezuma
Copeland
Bucklin
Coats
Ryus
Sublette
Satanta
Bloom
Minneola
Belvidere
Fowler
Sun City
Moscow
Wilmore
Lexington
Plains
Meade
Coldwater
Deerhead
Kismet
Hugoton
Rolla
Ashland
Sitka
Protection
Buttermilk
Liberal
Englewood
Tyrone
Willard
Lookout
Hough
Hooker
Baker
Turpin
Floris
Forgan
Mocane
Gate
Beaver
Rosston
Buffalo
Selman
Freedom
Optima
Adams
Laverne
Guymon
Avard
Hardesty
Balko
Elmwood
Slapout
May
Fort Supply
Waynoka
Goodwell
Logan
Gatesby
Booker
Darrouzett
Follett
Woodward
Mooreland
Perryton
Fargo
Farnsworth
Gruver
Waka
Gage
Shattuck
Sharon
Mutual
Spearman
Lipscomb
Chester
Higgins
Arnett
Harmon
Vici
Seiling
Morse
Taloga
Glazier
Camargo
Trail
Pringle
Canadian
Leedey
Stinnett
Plemons
Durham
Putnam
Roll
Miami

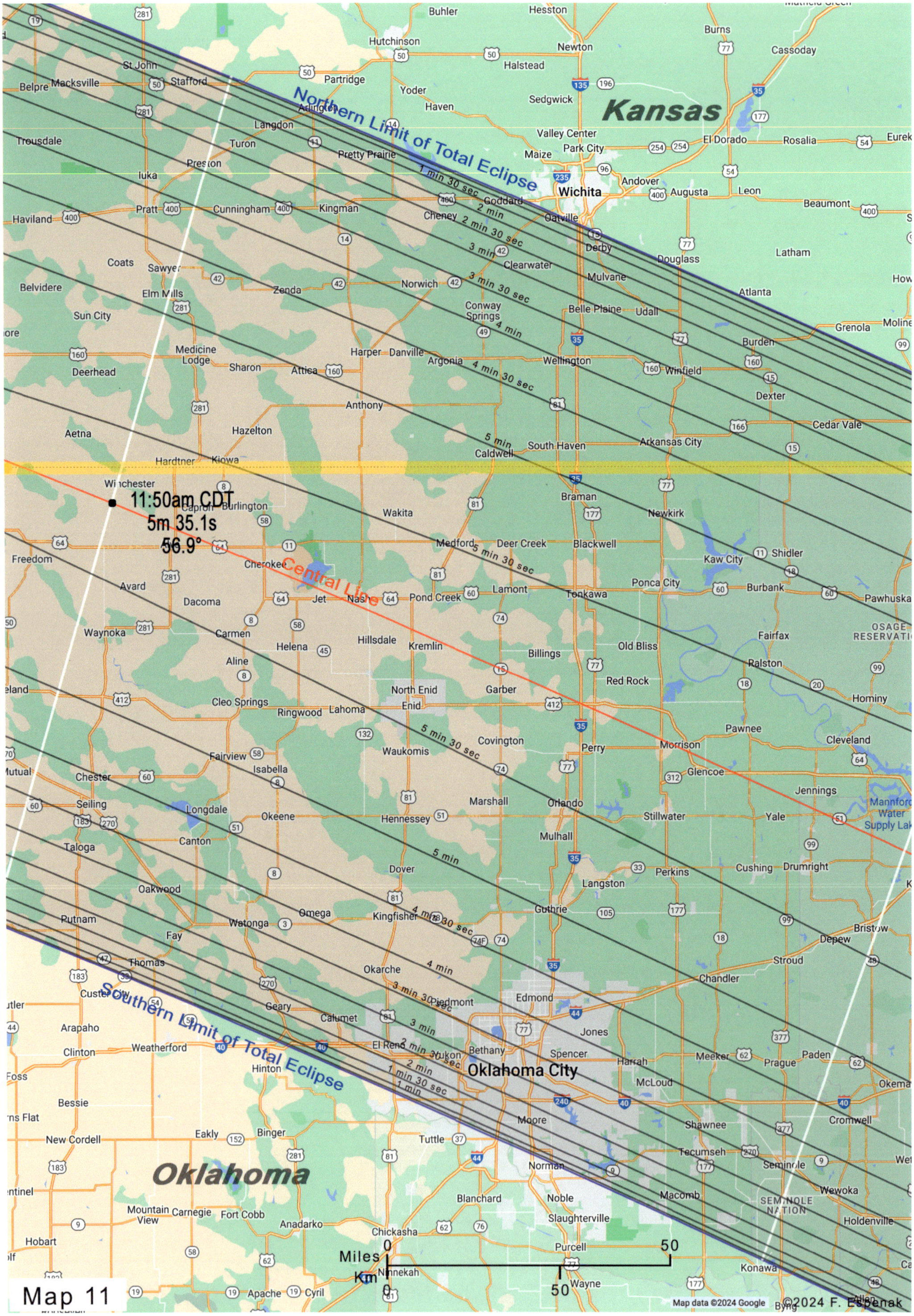
Kansas
Oklahoma
Northern Limit of Total Eclipse
Southern Limit of Total Eclipse
Central Line
11:50am CDT
5m 35.1s
56.9°
1 min 30 sec
2 min
2 min 30 sec
3 min
3 min 30 sec
4 min
4 min 30 sec
5 min
5 min 30 sec
5 min 30 sec
5 min
4 min 30 sec
4 min
3 min 30 sec
3 min
2 min 30 sec
2 min
1 min 30 sec
1 min
Wichita
Oklahoma City
Hutchinson
Buhler
Hesston
Newton
Halstead
Burns
Cassoday
St John
Stafford
Macksville
Belpre
Partridge
Yoder
Haven
Sedgwick
Valley Center
Park City
Maize
El Dorado
Rosalia
Arlington
Langdon
Turon
Trousdale
Preston
Pretty Prairie
Iuka
Pratt
Cunningham
Kingman
Goddard
Cheney
Andover
Augusta
Leon
Beaumont
Haviland
Oatville
Derby
Douglass
Latham
Coats
Sawyer
Clearwater
Mulvane
Elm Mills
Zenda
Norwich
Conway Springs
Belle Plaine
Udall
Atlanta
Belvidere
Sun City
Medicine Lodge
Sharon
Attica
Harper
Danville
Argonia
Wellington
Winfield
Burden
Grenola
Moline
Deerhead
Dexter
Anthony
Aetna
Hazelton
Caldwell
South Haven
Arkansas City
Cedar Vale
Hardtner
Kiowa
Winchester
Capron
Burlington
Braman
Newkirk
Wakita
Medford
Deer Creek
Blackwell
Freedom
Cherokee
Kaw City
Shidler
Avard
Dacoma
Jet
Nash
Pond Creek
Lamont
Tonkawa
Ponca City
Burbank
Pawhuska
Waynoka
Carmen
Helena
Hillsdale
Kremlin
Billings
Old Bliss
Fairfax
OSAGE RESERVATION
Aline
Red Rock
Ralston
North Enid
Enid
Garber
Cleo Springs
Ringwood
Lahoma
Hominy
Covington
Perry
Morrison
Pawnee
Cleveland
Waukomis
Fairview
Isabella
Glencoe
Chester
Mutual
Seiling
Marshall
Orlando
Jennings
Longdale
Okeene
Hennessey
Stillwater
Yale
Mannford
Taloga
Canton
Mulhall
Dover
Langston
Perkins
Cushing
Drumright
Oakwood
Kingfisher
Guthrie
Putnam
Omega
Watonga
Fay
Bristow
Depew
Thomas
Okarche
Stroud
Chandler
Custer City
Piedmont
Edmond
Arapaho
Geary
Calumet
Jones
Clinton
Weatherford
El Reno
Yukon
Bethany
Spencer
Harrah
Meeker
Prague
Paden
Hinton
McLoud
Foss
Okemah
Bessie
Moore
Shawnee
Cromwell
New Cordell
Eakly
Binger
Tuttle
Tecumseh
Seminole
Norman
Wewoka
Macomb
SEMINOLE NATION
Blanchard
Noble
Slaughterville
Holdenville
Mountain View
Carnegie
Fort Cobb
Anadarko
Chickasha
Purcell
Hobart
Ninnekah
Wayne
Konawa
Apache
Cyril
Byng
Miles
Km
0
50
0
50
Map 11
Map data ©2024 Google
©2024 F. Espenak

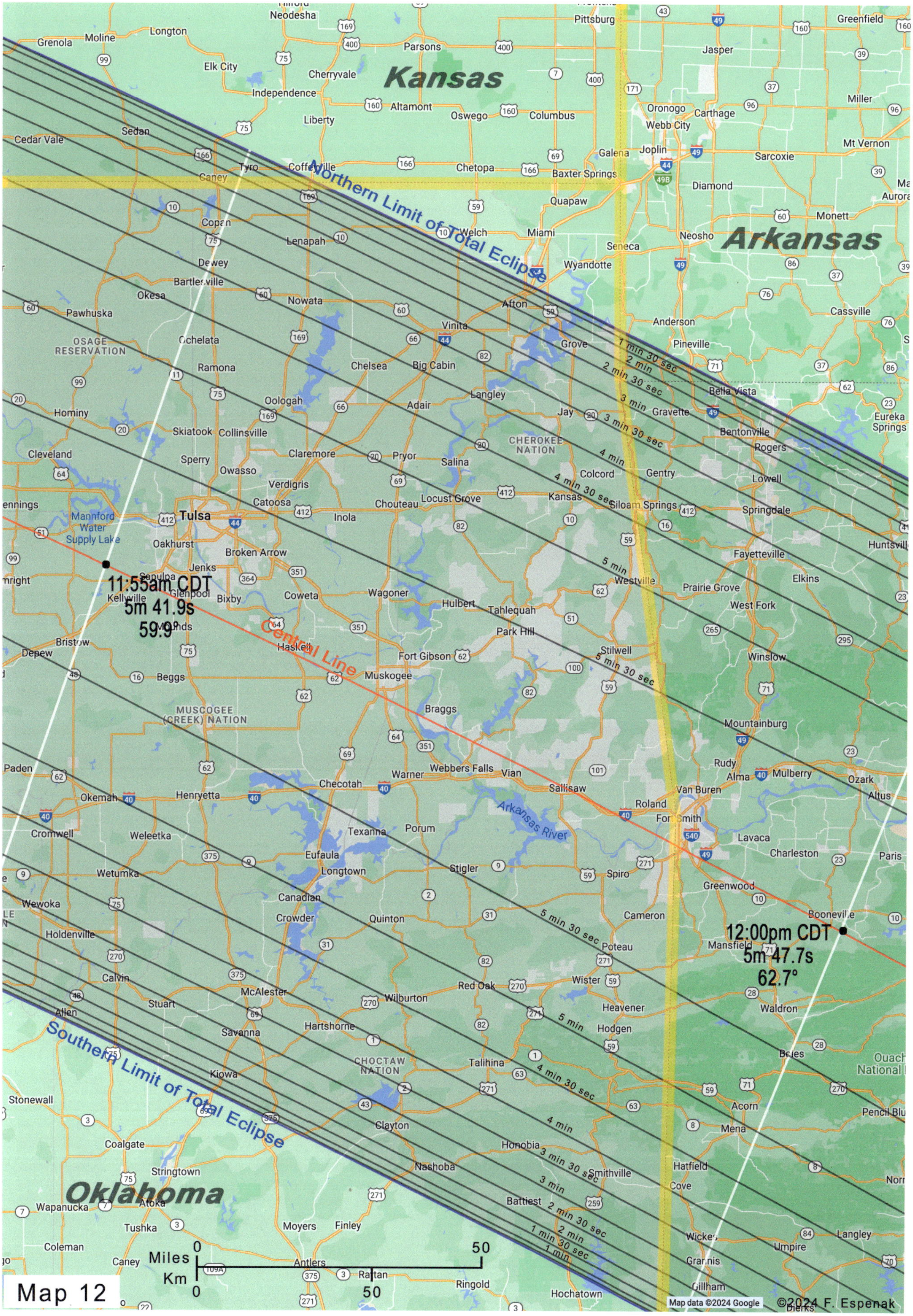
Kansas
Arkansas
Oklahoma
Northern Limit of Total Eclipse
Southern Limit of Total Eclipse
Central Line
11:55am CDT
5m 41.9s
59.9°
12:00pm CDT
5m 47.7s
62.7°
1 min 30 sec
2 min
2 min 30 sec
3 min
3 min 30 sec
4 min
4 min 30 sec
5 min
5 min 30 sec
Tulsa
Muskogee
Fort Smith
Fayetteville
Joplin
McAlester
Arkansas River
OSAGE RESERVATION
CHEROKEE NATION
MUSCOGEE (CREEK) NATION
CHOCTAW NATION
Miles 0 50
Km 0 50
Map 12
Map data ©2024 Google
©2024 F. Espenak

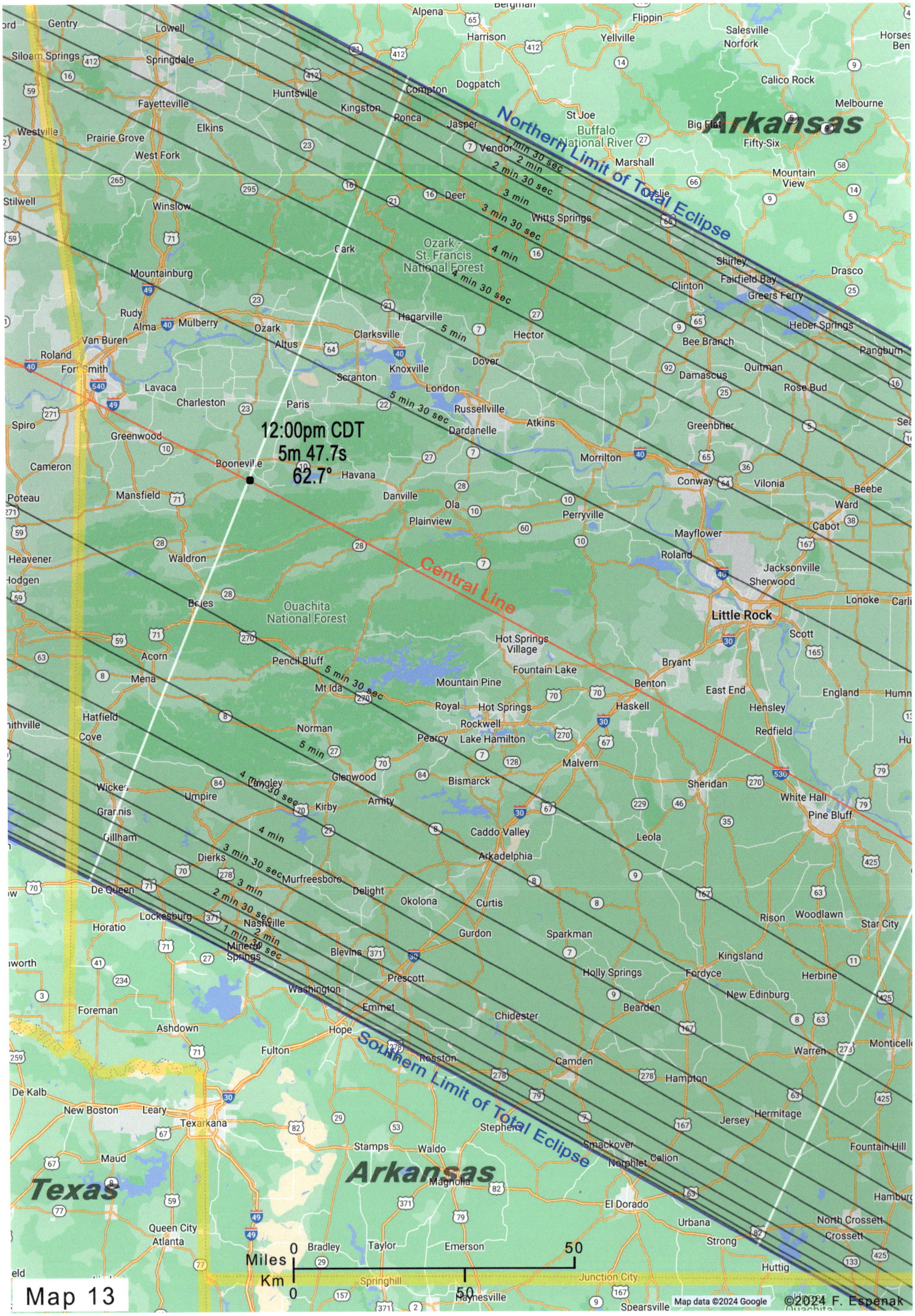
Northern Limit of Total Eclipse
Southern Limit of Total Eclipse
Central Line
12:00pm CDT
5m 47.7s
62.7°
1 min 30 sec
2 min
2 min 30 sec
3 min
3 min 30 sec
4 min
4 min 30 sec
5 min
5 min 30 sec
Arkansas
Texas
Ozark-St. Francis National Forest
Ouachita National Forest
Buffalo National River
Fort Smith
Fayetteville
Springdale
Harrison
Clarksville
Russellville
Dardanelle
Booneville
Paris
Conway
Little Rock
Hot Springs
Benton
Pine Bluff
Arkadelphia
De Queen
Texarkana
Camden
El Dorado
Heber Springs
Miles 0 50
Km 0 50
Map 13
Map data ©2024 Google
©2024 F. Espenak

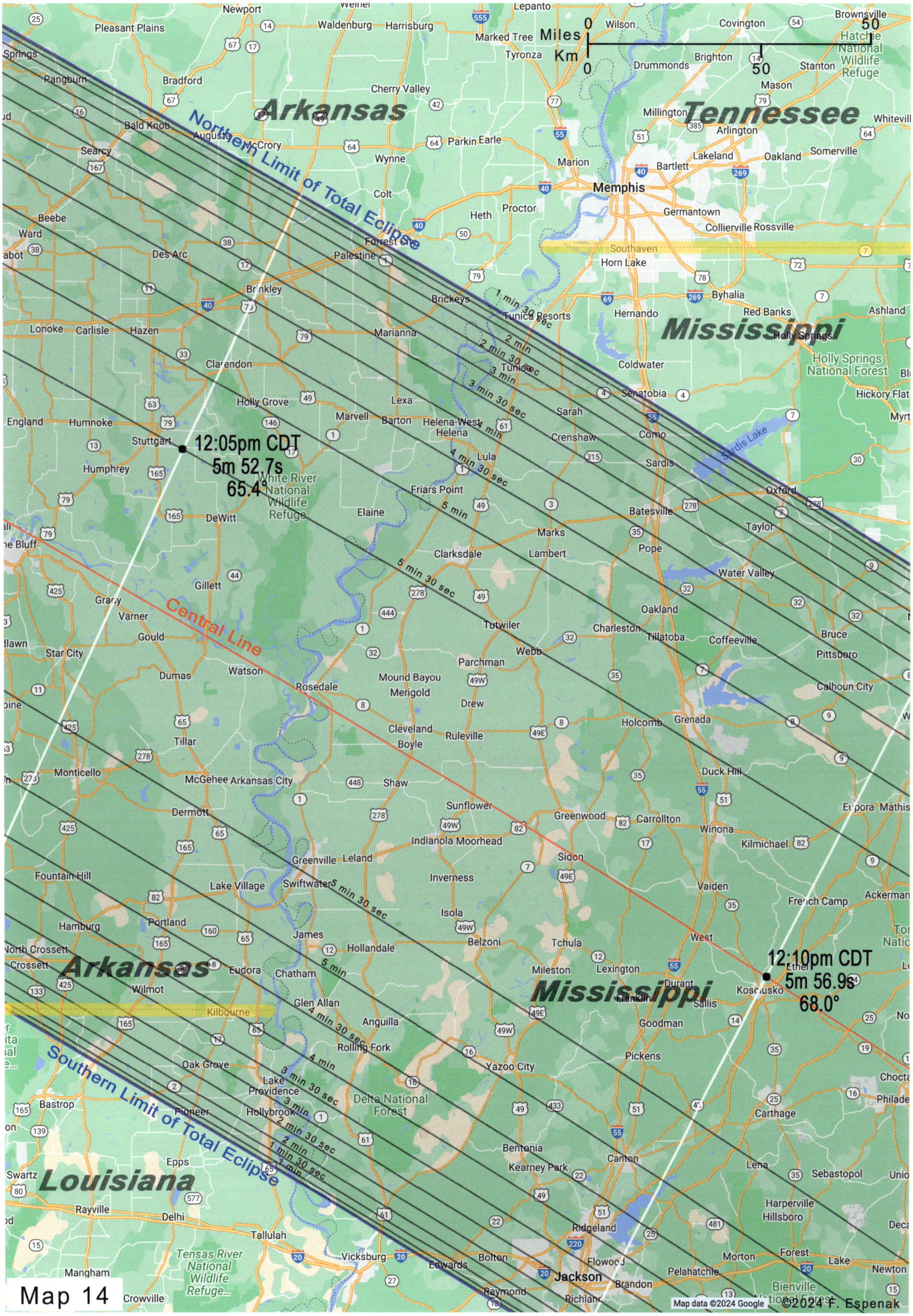
Arkansas
Tennessee
Mississippi
Arkansas
Mississippi
Louisiana
Miles 0 50
Km 0 50
Northern Limit of Total Eclipse
Central Line
Southern Limit of Total Eclipse
12:05pm CDT
5m 52.7s
65.4°
12:10pm CDT
5m 56.9s
68.0°
1 min 30 sec
2 min
2 min 30 sec
3 min
3 min 30 sec
4 min
4 min 30 sec
5 min
5 min 30 sec
Memphis
Jackson
Map data ©2024 Google
©2024 F. Espenak
Map 14

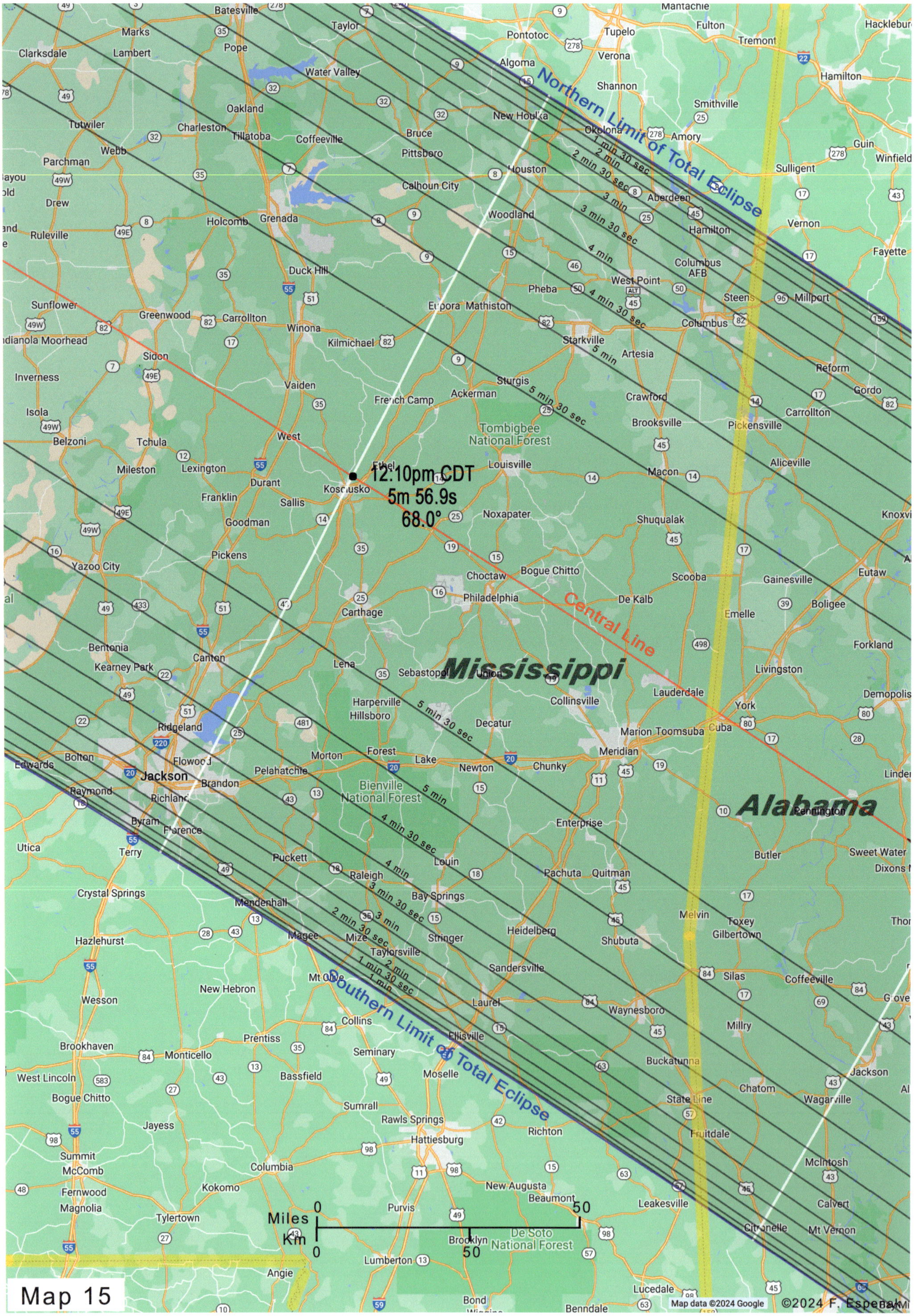
Northern Limit of Total Eclipse
1 min 30 sec
2 min
2 min 30 sec
3 min
3 min 30 sec
4 min
4 min 30 sec
5 min
5 min 30 sec
12:10pm CDT
5m 56.9s
68.0°
Central Line
Mississippi
Alabama
5 min 30 sec
5 min
4 min 30 sec
4 min
3 min 30 sec
3 min
2 min 30 sec
2 min
1 min 30 sec
1 min
Southern Limit of Total Eclipse
Miles 0 50
Km 0 50
Map 15
Map data ©2024 Google
©2024 F. Espenak

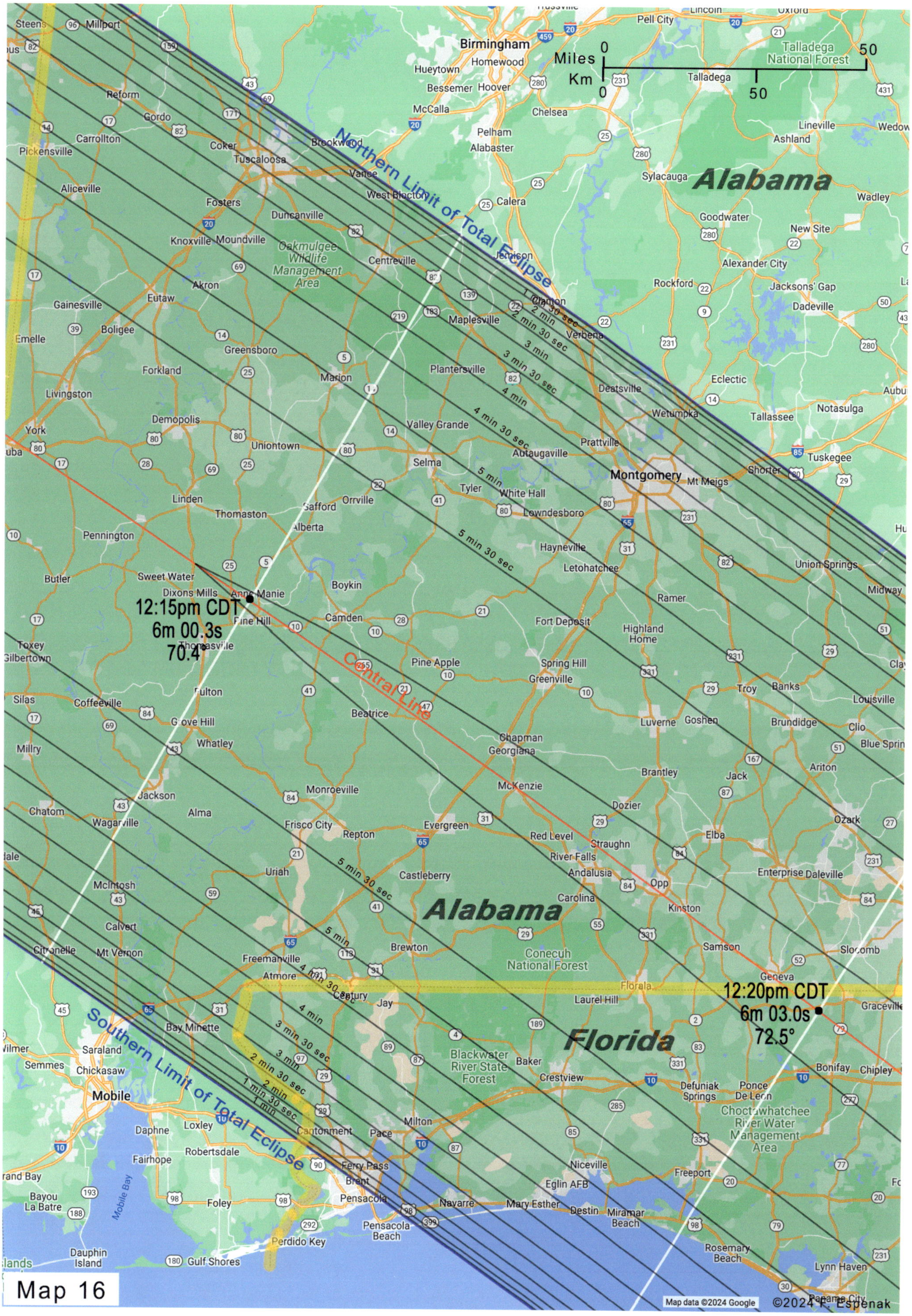
Map 16
Northern Limit of Total Eclipse
Southern Limit of Total Eclipse
Central Line
12:15pm CDT
6m 00.3s
70.4°
12:20pm CDT
6m 03.0s
72.5°
1 min
1 min 30 sec
2 min
2 min 30 sec
3 min
3 min 30 sec
4 min
4 min 30 sec
5 min
5 min 30 sec
Miles
Km
0
50
Alabama
Florida
Birmingham
Tuscaloosa
Montgomery
Mobile
Pensacola
Selma
Oakmulgee Wildlife Management Area
Conecuh National Forest
Blackwater River State Forest
Choctawhatchee River Water Management Area
Talladega National Forest
Map data ©2024 Google
©2024 F. Espenak

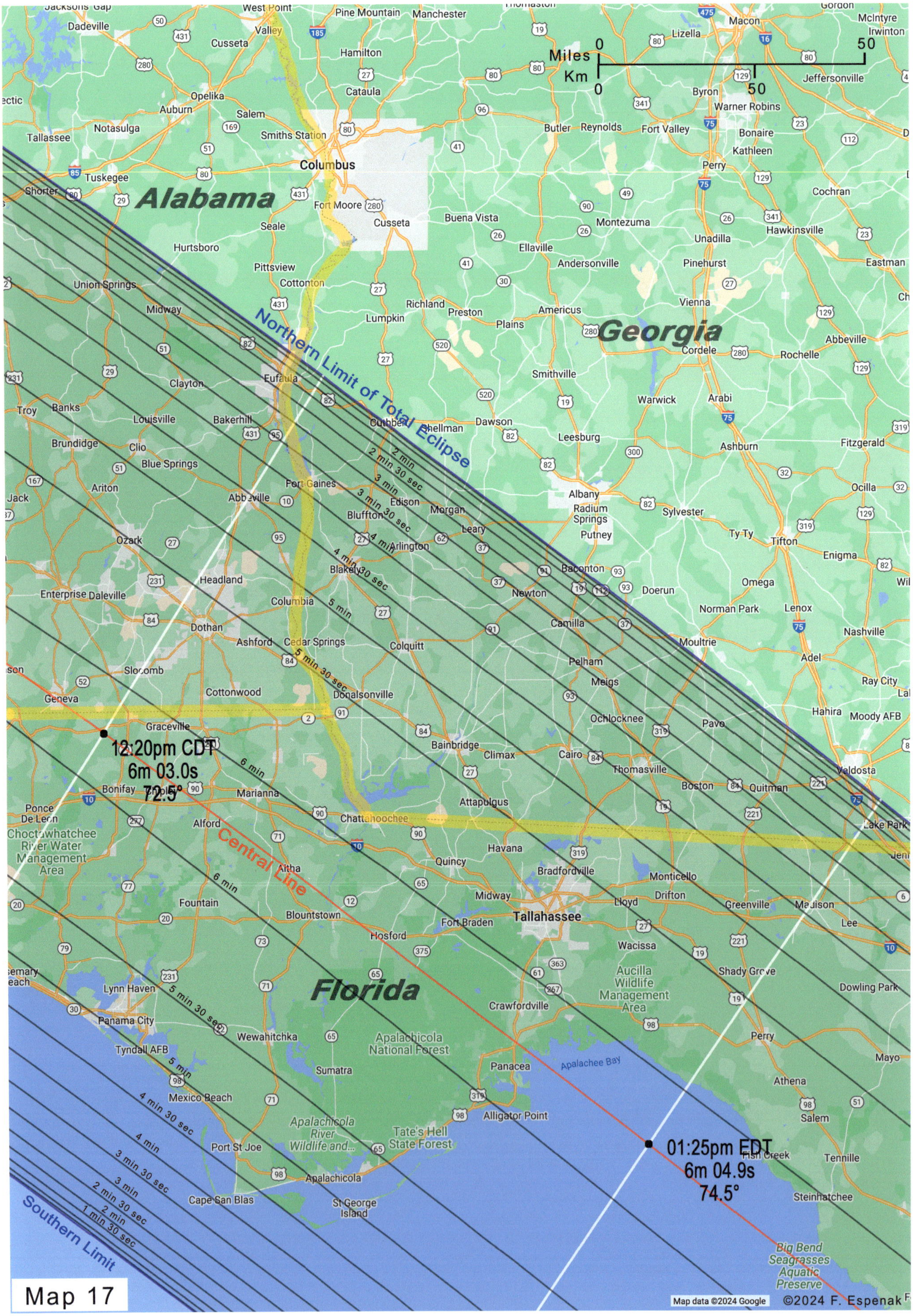
Miles 0 50
Km 0 50
Alabama
Georgia
Florida
Northern Limit of Total Eclipse
Central Line
Southern Limit
2 min
2 min 30 sec
3 min
3 min 30 sec
4 min
4 min 30 sec
5 min
5 min 30 sec
6 min
12:20pm CDT
6m 03.0s
72.5°
01:25pm EDT
6m 04.9s
74.5°
Columbus
Macon
Warner Robins
Albany
Valdosta
Tallahassee
Dothan
Panama City
Apalachee Bay
Map 17
Map data ©2024 Google
©2024 F. Espenak

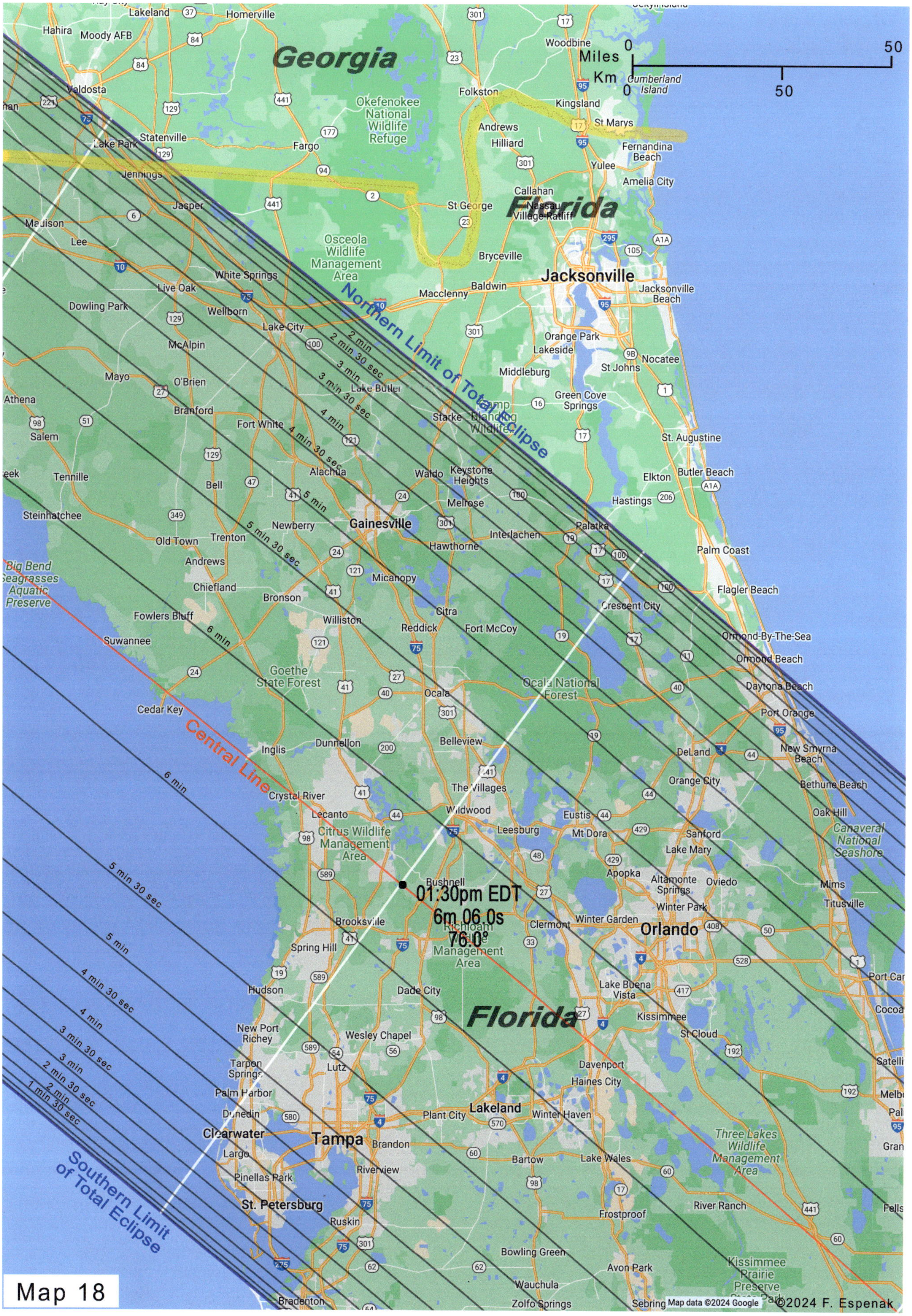
Georgia
Florida
Miles 0 50
Km 0 50
Northern Limit of Total Eclipse
2 min
2 min 30 sec
3 min
3 min 30 sec
4 min
4 min 30 sec
5 min
5 min 30 sec
6 min
Central Line
6 min
5 min 30 sec
5 min
4 min 30 sec
4 min
3 min 30 sec
3 min
2 min 30 sec
2 min
1 min 30 sec
Southern Limit of Total Eclipse
01:30pm EDT
6m 06.0s
76.0°
Valdosta
Jacksonville
Gainesville
Ocala
Orlando
Tampa
St. Petersburg
Lakeland
Clearwater
Daytona Beach
Map 18
Map data ©2024 Google
©2024 F. Espenak

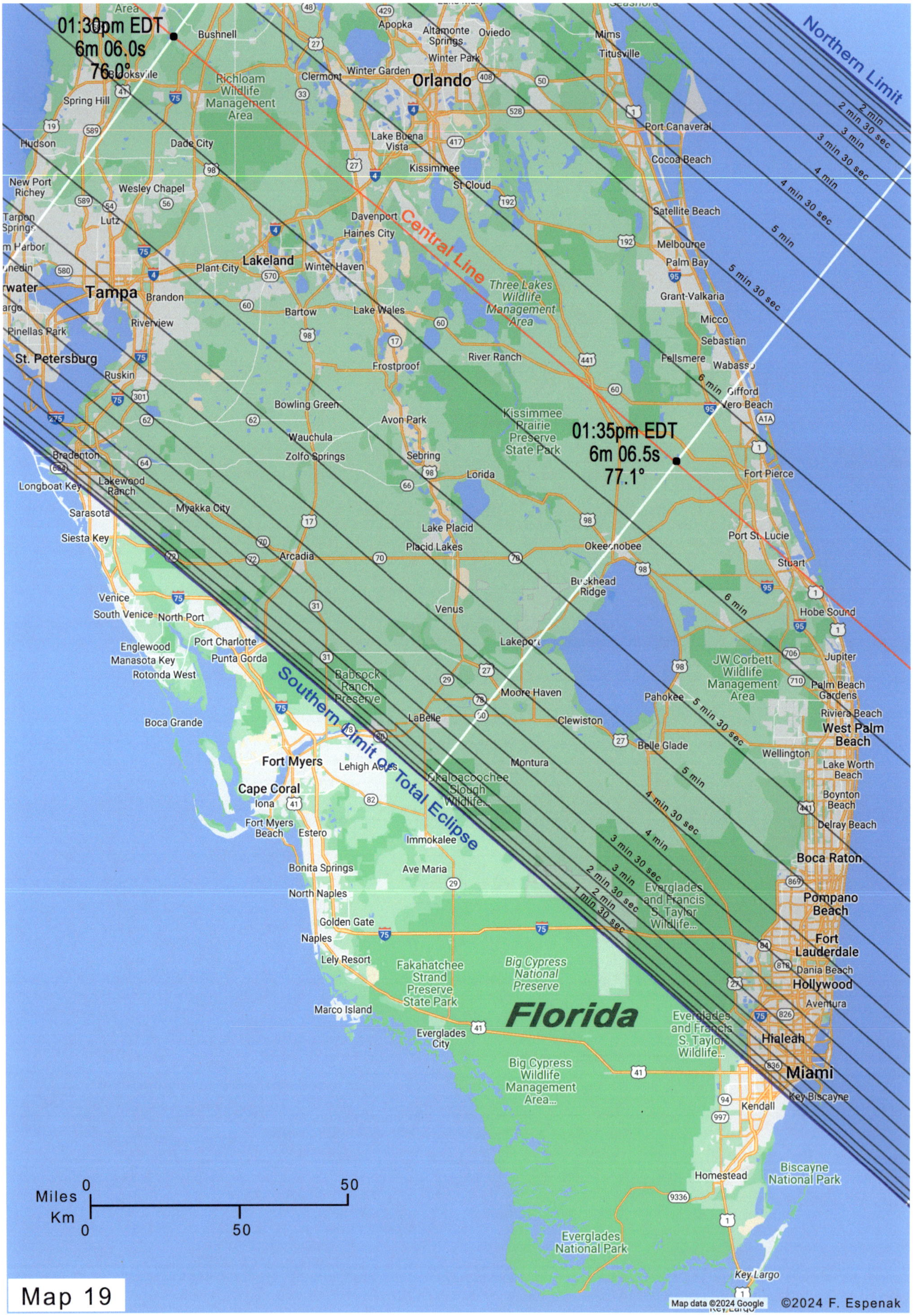
01:30pm EDT
6m 06.0s
76.0°
01:35pm EDT
6m 06.5s
77.1°
Northern Limit
Central Line
Southern Limit of Total Eclipse
2 min
2 min 30 sec
3 min
3 min 30 sec
4 min
4 min 30 sec
5 min
5 min 30 sec
6 min
1 min 30 sec
Florida
Orlando
Tampa
Lakeland
St. Petersburg
Fort Myers
Cape Coral
Miami
Hialeah
Fort Lauderdale
West Palm Beach
Boca Raton
Miles
Km
0
50
Map data ©2024 Google
©2024 F. Espenak
Map 19

Eclipse Guides from AstroPixels Publishing

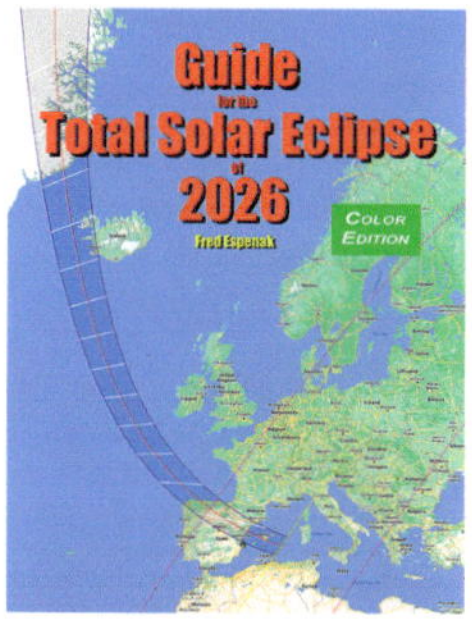

Guide for the Total Solar Eclipse of 2026

The total solar eclipse of August 12, 2026 will be visible from a path across northern Russia, the Arctic, Greenland, Iceland, and Spain. The **Guide for the Total Solar Eclipse of 2026** contains 14 full-page maps of the path. The large scale shows roads, towns and cities, rivers and lakes.

Tables for hundreds of cities provide times each eclipse phase of the along with the eclipse magnitude, duration, and the Sun's altitude. A climatology study identifies areas along the eclipse path where the highest probability of favorable weather may be found.

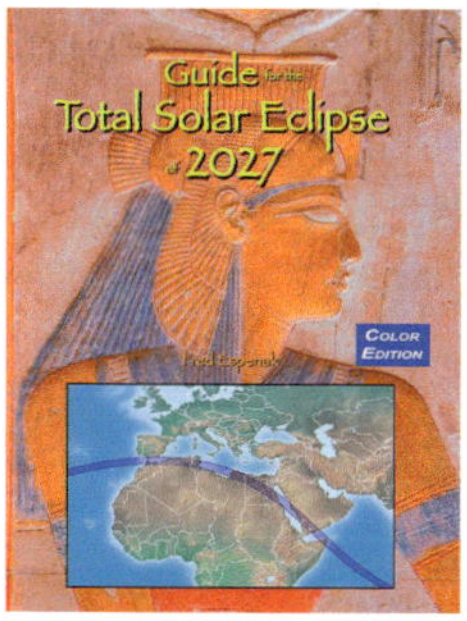

Guide for the Total Solar Eclipse of 2027

The total solar eclipse of August 2, 2027 will be visible from a path across Spain, Morocco, Algeria, Tunisia, Libya, Egypt, Sudan, Saudia Arabia, Yemen, and Somalia. The **Guide for the Total Solar Eclipse of 2027** contains 20 full-page maps of the path.

Tables for hundreds of cities provide times for each eclipse phase along with the eclipse magnitude, duration and the Sun's altitude. Cloud statistics identify areas where the highest probability of clear sky may be found.

Guide for the Total Solar Eclipse of 2028

The total solar eclipse of July 22, 2028 will be visible from a path across Australia and New Zealand. The **Guide for the Total Solar Eclipse of 2028** contains 19 full-page maps of the path of totality.

Tables for over one hundred cities provide times for each eclipse phase with the duration of totality and the Sun's altitude. Cloud statistics identify areas where the highest probability of clear sky may be found. A preview is given for total eclipses through 2039.

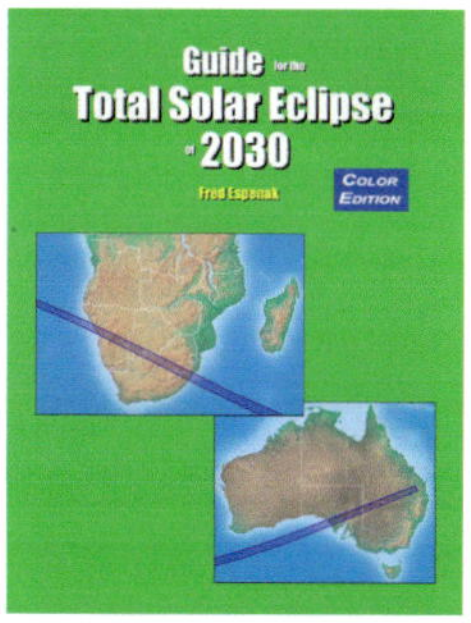

Guide for the Total Solar Eclipse of 2030

The total solar eclipse of November 25, 2030 will be visible from a path across Africa and Australia. The **Guide for the Total Solar Eclipse of 2028** contains 17 full-page maps of the path of totality.

Tables for over one hundred cities provide times for each eclipse phase with the duration of totality and the Sun's altitude. Cloud statistics identify areas where the highest probability of clear sky may be found. A preview is given for total eclipses through 2039.

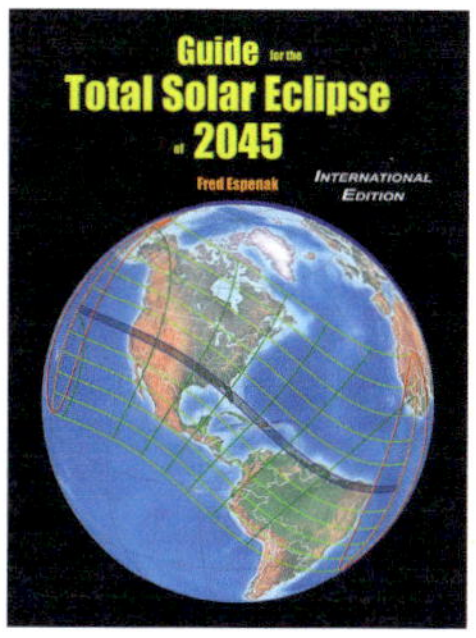

Guide for the Total Solar Eclipse of 2045

The total solar eclipse of August 12, 2045 will be visible from a path across the USA, the Caribbean, and South America. The **Guide for the Total Solar Eclipse of 2045** contains 37 full-page maps of the path of totality.

Tables for hundreds cities provide times for each eclipse phase with the duration of totality, eclipse magnitude, and the Sun's altitude. Cloud statistics identify areas where the highest probability of clear sky may be found.

For more information on each *Guide*, including sample pages of each, visit:

astropixels.com/pubs/guides.html

The *Guides* may be ordered from ***Amazon***

Eclipse Books from AstroPixels Publishing

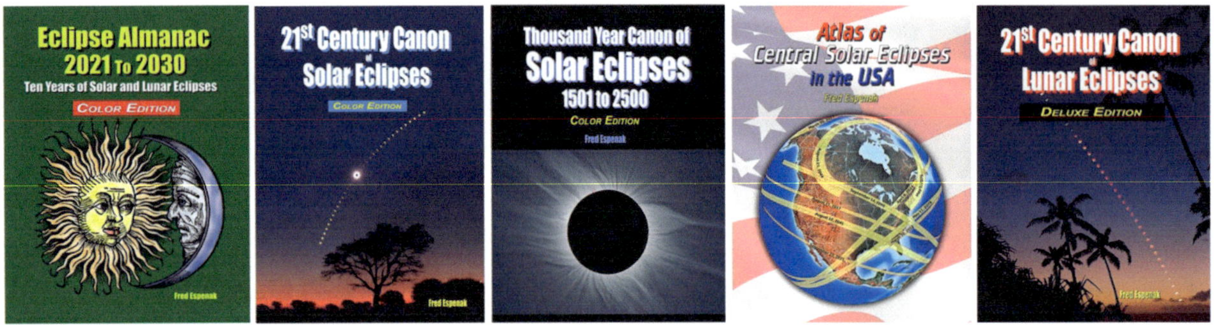

Eclipse Almanac – 2021 to 2030

A concise reference for every eclipse of the Sun and the Moon over a 10-year period. It identifies when and where each event will be seen with maps for each. Additional details about each eclipse are included. Other volumes in this series include the decades 2032 to 2040, 2041 to 2050, 2051 to 2060, and 2061 to 2070.

21st Century Canon of Solar Eclipses

The complete guide to every solar eclipse occurring from 2001 tom 2100 (224 eclipses in all). It includes information and maps for all total, annular, hybrid, and partial eclipses. A special world atlas shows detailed, full page maps of all central eclipse paths (total, annular, and hybrid).

Thousand Year Canon of Solar Eclipses — 1501 to 2500

Contains maps and data for each of the 2,389 solar eclipses occurring over the ten-century period centered on the present era. The maps are arranged twelve to a page at an image scale allowing the determination of eclipse visibility from any location on Earth. A comprehensive catalog lists the essential characteristics of each eclipse.

Atlas of Central Solar Eclipses in the USA

When was the last total eclipse through the USA and when is the next? How often do they happen? What total eclipse tracks passed across the USA during the 17th, 18th, and 19th centuries, etc., and what states did they include? And how often is a total solar eclipse visible from each of the 50 states? The Atlas of Central Solar Eclipses in the USA answers all these questions and more, with hundreds of maps and tables.

21st Century Canon of Lunar Eclipses

The complete guide to every lunar eclipse occurring from 2001 tom 2100 (228 eclipses in all). It includes information and maps for all total, partial, and penumbral eclipses. The predictions use a new model for Earth's elliptical shadows.

Thousand Year Canon of Lunar Eclipses — 1501 to 2500

Contains maps and data for each of the 2,424 lunar eclipses occurring over the ten-century period centered on the present era. The maps are arranged twelve to a page at an image scale allowing the determination of eclipse visibility from any location on Earth. A comprehensive catalog lists the essential characteristics of each eclipse.

All books are available in two editions: 1) Black and White, and 2) Color.

For more information including sample pages of each, visit:

astropixels.com/pubs/guides.html

Each *Book* may be ordered from ***Amazon***

Made in the USA
Monee, IL
05 November 2026

67383126R00026